GROSS
A Paramedic Story

By

Craig A Mills
and
Justin J Miller

ISBN 978-0-6151-6454-0

This book reflects our experiences in our own words, as accurately and honestly as possible. It does
not purport to reflect the views or opinions of the Ambulance Service in which we are presently
employed and they have not officially approved or disapproved of its content.

We cannot identify patients or locations and may change certain facts in order to protect patient
confidentiality and we hope you understand why this is necessary. This is a work based on the true
events and experiences we as paramedics have encountered, all events and facts are not completely
accurate which is part of the protection of patient confidentiality.

The opinions expressed in this book are our own and do not reflect upon anybody but ourselves.
Those opinions are subject to change and may have already changed by the time of this printing, but
they do reflect the nature of this job and are a window into Paramedicine. It is but a small glimpse into
who we are as paramedics and the experiences that so alter and affect us.

Acknowledgements

This book was inspired by our passion to be paramedics. It was encouraged by our students and partners, who saw the things we saw and heard the stories that resulted from our experiences. We would like to thank all those wonderful people for their understanding and encouragement.

Yet we wouldn't want to contribute to that "paramedic ego," which is so over-inflated and out of proportion. (A little sarcasm here, something we tend to develop as we respond to people's appreciation, or rather, lack of appreciation.)

You know what?

Forget such false humility and subservient propaganda.

We need to be thanked! We need to be recognized and we need to be respected.

No, not for us; we don't need such affirmation.

No, the ones who would recognize us, they need to give us our place, to give us our due, that they should recognize our struggle and effort in order that a little respect would be mutual. Respect is always, and must always be a two-way street.

If nobody wants to give us this respect, then let them come out here and do this job.

Funny thing is we don't consider it a job. That is the difference. We consider it a calling.

So, we are going to thank all those who have contributed and have been a part of this grand adventure, this grand calling. So here goes:

Thank you, John and John, Holly and Rick, Scott and Scott, Justin and Jason, and Jason. Thank you, Mark and Doug and Larry. Thank you, Dave and Jeanna and Ryan. Thank you, Jeff and Randy, and Bob and Larry, and Dawn and Jan, and Valerie.

Thank you, Peter and Grif and Laurie and Twyla, Greg and Terri. Thank you, Dawn and Mike. Thank you, Mike and Jennifer.

Thank you, Victor and Frank. Thank you, Henry and Larry. Thank you, Matt and Pat and Mike. Thank you, Michelle and Pam and Steve and Mike. Thank you, Steve and Robin.

Thank you, DJ and Lisa and Rebecca and Joe and Mitch. Thank you, Sam, Jamie and Tammy, Kim and Larry. Thank you Chad and 'the Chad'. Thank you, Jeff and Mike and Dawn and Lisa and Oliver and Amy.

And special thanks to Craig's wife and daughter, (Gloria and Ruth respectively) who have helped us type and edit and retype and reedit this undertaking. And thank you to Sue N. and Pat W. for your contributions. (Midsummer's night nightmare and Mr. Spoon, respectively). Thank you Rebecca for 'fingers and toes'. Thanks Sarah A. for your help with grammar and punctuation. Thank you to our families, who have remained steadfast during our readings.

And if you have worked with us and we did not mention you, then thank you! To the unnamed medic who has been a partner and, at times, a friend. We give you the respect and honor that is due. Remember our dementia and forgive us.

Memory loss is an essential part of this job; it is our safety mechanism that keeps us sane.

A special thanks to Jeff Selby for designing and editing our book cover. Thanks! AIW Creative, for your contribution.

Http://www.aiwcreative.com

Table of Contents

INTRODUCTION
GROSS A Paramedic Story

A Case for Insanity

To be, or not to be: that is the question: Whether 'tis nobler in the mind to suffer the slings and arrows of outrageous fortune, Or to take arms against a sea of troubles, And by opposing end them? To die: to sleep; No more; and by a sleep to say we end The heart-ache and the thousand natural shocks That flesh is heir to, 'tis a consummation Devoutly to be wish'd. To die, to sleep; To sleep: perchance to dream: ay, there's the rub; For in that sleep of death what dreams may come When we have shuffled off this mortal coil, Must give us pause: there's the respect That makes calamity of so long life;

Hamlet

-William Shakespeare

Here we sit in all our glory. "To be or not to be?" That is our outrageous fortune and our question.

Pressed uniforms, now wrinkled from sitting in our cramped

metal box, and shiny combat boots, like men waiting to go to war. Curled up in our metal box on wheels, the harbinger of hope and death, depending on who you are, all wrapped up in one – AN AMBULANCE. Soon to be called a mobile home because it seems I practically live in this box. All cramped up in the fetal position, preparing me for a nursing home. Sleep depravation and madness only around the corner.

For in that sleep, to die, to sleep and dream... what dreams may come?

No, we can't sleep, let alone dream. We must be ever vigilant.

We want to tell you about our insanity – about our reality, but we find that words are not quite enough. Perception is not necessarily reality, but it might be the reality of the perceiver. Your reality is definitely not always true, just wait; you will see. For in these stories we will show you our reality, and it is definitely unreal.

No more; and by a sleep to say we end the heart-ache and the thousand natural shocks that flesh is heir to, that is the unreal, (our natural shocks that flesh is heir to – **"Copy a code 3" blares over the radio, disturbing us from our sleep which we are not suppose to be doing**- the world of the paramedic.

We thought about using pictures, but they are too gross (hence the name of our book), so we are going to use the most descriptive words we can find.

Together, as we recollect our stories, hopefully we will give you a glimpse into our insane, bizarre and crazy world. By the way, this happens to be your world. You just don't realize it yet. We may shatter your reality so please forgive us this indiscretion.

Day in and day out, responding to people's cries for help, we scurry like mice on a sinking ship, partially meeting people's needs, partially persuading ourselves that we make a difference.

Okay, maybe it's just somebody being held at gunpoint by the police. (*This is happening as I sit here and write this.*)

Someone is ready to kill somebody with a butter knife. Police are called and we are told to "stage" out of the area. We wouldn't want to put ourselves in danger of being cut by a butter knife.

Then we are canceled. No need. No, just a false alarm.

So they took the butter knife away from the knife-wielding killer. It was only a butter knife, but someone was going to get really hurt for stealing his shopping cart (a glimpse into our insanity).

The insanity is just beginning today, but I digress, back to the introduction.

My Reason and Disclaimer

To all you EMS workers out there; you know whom you are, those who might read this book. It is to you we write some of our stories; the other ones, I don't know who they are for, maybe the curious ones.

This book addresses the true underbelly of the EMS and maybe even the backside too, so if it offends you, I'm sorry, so remain in your perception of your reality. Some day, reality will hit you square in the face and then you will understand.

Realize this: the EMS system is really, only an illusion, *smoke and mirrors*, placed there by well-intentioned government folk to make us feel like they're doing something. They want to say they made a difference, and to make up for that law that says we have to transport everybody who wants to ride in our metal box. Their good intentions only created another government bureaucracy with all its red tape and mismanagement. The system has been set up to overload and implode. Anyone who picks up the phone and dials 911 has an emergency by virtue of the act itself. This is the flaw in the system.

So I hope you look good in the mirror? Remember the smoke and mirrors I mentioned earlier. No, add more smoke, that's better; It's more fitting since you will be around a lot of smoke, literally and figuratively.

Remember these words and when the shit hits the fan and your support system reveals its ugly head, you will at least know you have been warned. Maybe you will retain some of your sanity. So don't get caught humming in the fan, when the shit hits the fan that is. (I love humming in the fan its great therapy.)

If you see yourself in any of our stories, well it's merely coincidence, and I'm sorry you found yourself immortalized in EMS legend. (I was told some of ours stories are stuff of EMS urban legend).We tried to be creative with the names and have embellished some of the stories to give it the necessary flavor, so we hope you enjoy.

We would like to invite you on our rollercoaster ride. There will be ups and there will be downs. You will feel the adrenaline rush, the nausea and the laughter. The emotional G-forces may be pretty intense so hang on tight and get your "barf bag" out and your Zofran or Phenergan (anti-nausea medication). Come with us, and enjoy the ride.

Chapter I

House of the Living Dead

So, let's go to the house of the living dead. I know, an oxymoron, but bear with me. You will come to understand about the living dead as we tell you our stories.

Well, some are living, but mostly dead.

There are several levels of "dead," you will soon discover.

They don't teach you this in school. They don't tell you that it's part of your job to "triage" (sort) the dead. What level of dead are they?

There are three basic levels of dead. There is partly dead – those who we try to resuscitate with all our efforts. Then there are those who are mostly dead – those who we call the medical examiner so that he can come check out to see how dead they really are, and then there are the DRT's – dead right there. These are the dead who we know should have probably died a long time ago, and have finally succumbed to the ravages of time. These are the ones we go through their pockets and look for loose change (just kidding) or an ID.

Wait, I thought we were supposed to save people!? I thought we were responding to a 911 call for help. How do the dead call 911 anyway?

So, maybe they are partly dead and we can pull them back from the brink of death. No! We're too late; some idiot wouldn't get out of our way. Remember that traffic jam we had to go through to get to you. Two minutes too late. We were sitting on our asses too long. No, we were too late for such heroism. Angel of death, ambulance driver – HACK with a patch. We know the derogatory terms used to describe us and we even use them amongst ourselves, but please remember our struggle and maybe you will change those words to "savior" or "hero" as an act of kindness.

Let me tell you what I have learned from dead people (sometimes they're better teachers than those who profess to teach). Yes, they can teach you something.

Dead people are neither objective nor subjective. They are neither critical nor complimentary. They usually don't call you names and be abusive to you for being too late. They really can't carry on a conversation, but they sure can teach you a lot about life, another oxymoron. I will tell you, they are very good at posing. I have seen the praying dead man, the floating dead man, the standing dead man, and yes, even the dancing dead man. Ladies, don't feel left out; I've seen my share of dead women. In my experience though, women don't seem to pose as well as men when it comes to death.

Oh, yes, they all managed to call 911!

How is that?!

Well, those well-intentioned have actually called 911, because that is what they have been told to do. Get help, call 911!

The public is educated to the point of ambivalent ignorance. Taught how to do just enough (call 911) but not to do any more; sure they learn CPR, but most don't do it. The thought of putting your mouth on a dead stranger and vomit is a deterrent and Liability is the excuse! I guess we need to treat the lawyers, for the law to change. Maybe they would change the law if they had enough heart attacks or strokes, or at least rethink it? I know ranting won't change anything so I won't continue.

So, let us go to the house of the dead and learn about the living dead.

The Dancing Dead Man

Let me tell you about a dancing dead man. Yes, believe it or not dead men can dance.

A ninety-two year-old man tells his caregiver he is going to die today. He has come to live in a nursing home. He has outlived all his close relatives and he has nowhere else to go. We call this TMB – too many birthdays, a problem we all should have, but really don't want. As Benjamin Franklin once said, *"All would live long, but none would be old."*

Mr. Stiff is a thin frail man who has seen many years of hard labor, and is now skin and bones, literally. He probably weighs not much more than a hundred pounds all wet, and he is five foot ten.

With his experience, he probably knows more than the young caregiver does.

No, he definitely knows more than she does.

He knows he is ready to die. He is ready to dance.

She smiles her young cute smile with her teenage little mind thinking, "*Oh, how sweet.*"

"No, Mr. Stiff, you're not going to die today," she says in a comforting ambivalent tone.

Later, she comes into Mr. Stiff's room to help him to the toilet. She uses the portable toilet this time because Mr. Stiff doesn't seem as mobile today.

Mr. Stiff is sitting in his chair staring at her as she comes into his room, with his eyes wide open.

"Well, Mr. Stiff, put your hands up on my shoulders," but he just sits staring with that continuous stare.

She lifts his stiff arm up, places it on her shoulder and pulls Mr. Stiff up, rotating him to the porta-potty. As he sits, she hears him move his bowels as he stares at her with that same odd, blank expression. She lifts him from the toilet and cleans him.

"That's good, Mr. Stiff. Don't worry; I'll clean you up. I do this all the time."

"You seem awfully quiet today, Mr. Stiff?" she says as she

moves him back to his chair.

"You seem awfully cold. Let me get you a warm blanket." He still stares at her with his eyes wide open.

"See you at lunch," she says in a friendly tone, leaving to go to attend to another patient.

An hour later, another caregiver checks on Mr. Stiff and runs out to call 911 with her eyes wide open. She comes back and moves Mr. Stiff to the floor as he stares up at the ceiling with that same blank stare. He is so stiff; his knees won't stay down as she pushes on them. His head is up so she places a pillow under his head to make him comfortable.

Mr. Stiff is staring at nothing with his eyes still open. Poor Mr. Stiff... he is so stiff he can't lay flat.

I arrive and take out the pillow. (Now a procedure I invented called "The Pillow Test.")

Mr. Stiff is still sitting up with his knees bent, with no pulse, not breathing. He has rigor mortis. Mr. Stiff is dead, but at least he has a clean ass; by the way, dead men do crap when moved, and especially if you dance with them.

As I declare Mr. Stiff dead, at least two hours dead, the cute, sweet aide throws her hands up in the air and screams as she runs from the room. She must have realized she gave Mr. Stiff his last ass wipe and his last Foxtrot. She had never seen a dead man before, let alone danced with one. Mr. Stiff was her first dead man and her first dance with a dead man.

I guess there is always a first and a last dance.

The Praying Dead Man

A man drives up to the emergency ambulance bay in his Plymouth Caravan.

It's a hot summer day. No, it's a scorching summer from hell, one of the hottest on record.

"I think I have a dead man in my van!" he says.

"No, I didn't call 911! I just drove up to the door of the ER where the ambulances are parked and I need your help. This was the best I could do with a 'praying' dead man."

We look at him wondering if the heat has gotten to him.

"He's your emergency now," he tells us.

I open the door of the van and there kneels a young construction worker. A wave of odor assaults me as the heat rushes past me.

This heat wave really makes a man smell, especially a dead man.

How do you describe the smell of a dead man?

Take a deep breath… no… hold your breath. You will never forget 104 ° F death. It's rank, yet sweet. It makes your eyes water and it burns a memory into your olfactory nerve that you will wish you could forget. No matter how many times you blow your nose, that odor is still a pungent part of each breath. The only consolation is the smell will go away in about a day.

They used to say breathe through your mouth so it won't be as bad.

They lied to you!

You can taste the odor of death and it's not pleasant.

I know, just don't breathe, that's the best, unless of course you pass out, but then I guess you won't smell the odor of death.

I don't pass out while I hold my breath, as I take his pulse, which is absent so I know he has gone to meet his maker.

Do dead men pray?

I guess they do. Well, this one does. As I tap him on his shoulder, I notice the stiff deformed, distorted, bluish face. (They call this rigor mortis with dependent lividity – that bluish color.) He has a death grip on the back bench where he is kneeling in the van, praying.

He's dead. Rigor mortis dead! I declare him dead and tell security to call the police. I guess he didn't get his prayer answered, or maybe he did? He can't feel the awful heat outside, not any more anyway. He has assumed room temperature, and I think he has a fever of 104°F.

We have another call for help, so we have to run. We have to go so someone else will deal with the praying dead man. As we drive off, someone shouts, "He's not dead, get the stretcher, get him out, put an oxygen mask on him, he is kind of blue."

Our noses lied to us! No, that's just not right.

We get canceled at the bottom of the hill.

Good, our curiosity is piqued. We quickly turn around to see what we missed.

Maybe his prayers were answered, maybe he has risen from the dead? Let's go back and see the miracle of the praying dead man.

As we pull around, Mr. Miracle is still praying, but now he is on his back on a stretcher.

He kind of looks like a dog on its back with his hands and feet out stiff in the air and an oxygen mask on his face. He is still blue and the seat cushion of the van is imprinted on his face. His knees are still bent, so I think he would like to say grace. No, I'm mistaken. He is dead. He didn't get his prayers answered. It's still hotter than hell and he's still praying.

The doctor comes out and says, "Guys, come on! This man is dead." So much for protecting the crime scene; This *was* a crime scene, by the way, a mobile crime scene.

Goddamn this heat!

I thought I would get to see a miracle, but all I saw was a praying dead man that smelled. One who was dead, then alive and then dead again... I wish he would make up his mind.

The story finally came out about our praying dead man. He was a construction worker who got too hot and went to take a nap in his van. He was overcome by the heat yesterday and was found this afternoon by his co-worker. The one who was so kind to give him a ride to the hospital.

Baking for the last twenty-four hours, Mr. Miracle was a well-done construction worker, hence the smell. His death was caused by heatstroke and dehydration.

I don't think he will be going back to work, no matter how hard he prays.

The Floating Dead Man

"Unconscious, possible code 99!" blares over the radio. We are the closest to the call so we are dispatched posthaste: Time to go to work rescuing the dead.

As we walk into the home of an elderly couple, the wife says to me, "I was just out shopping. I was only gone for an hour. He had gone into the spa to have a relaxing morning. He looked so comfortable with his head back on the headrest and his arms up on the edge of the spa. His health hasn't been so good and the spa was so relaxing for him. I didn't think it would kill him. We keep the spa 'pretty warm'. That's how we like it."

Mr. Spa looks like a waterlogged board.

From the tits up, like a department store mannequin floating in the river; from the tits down, like a floppy fish – or a wet noodle.

We pulled Mr. Spa from his relaxing spa – "lobster legs" is a term that comes to mind, but at the moment I think it unprofessional to mention. As we lay his body carefully onto the deck with his arms pointing to the sky, but his legs (red lobster legs) flopping on the slippery deck, we confirm him relaxed to the point of being dead. Yes, Mr. Spa was relaxed to death. I think I would like to find my fate such as his. Most of us hope to die in our sleep, because we have seen the struggle for life and it often isn't pretty.

Throw a frog into boiling water and he will jump out, turn up the heat and he will boil to death.

Throw a man into a cold spa and he will jump out, but put a man into a hot spa and he will relax to death. Go figure; is that a parallel conundrum or what? Mr. Spa is finally relaxed, just six feet under. I would love to relax in a spa, but probably not that much. My question to the medical community is, "Where was the education that would have prevented this tragedy?" Maybe the spa company was negligent when they sold the elderly couple the spa? I think I remember reading the warning label and it does say to consult your doctor. Maybe they didn't consult their doctor?

Well, Mrs. Spa will have to enjoy the spa alone. I hope she consults her doctor, lest she join her husband in that great spa in the sky.

The Gardening Dead Man

"Code 99": Not many know the term in EMS for dead man. "Code 99" is the increasing heart rate, the sweat on the brow, for the rookie. For the gung-ho medic, it's the adrenaline rush, but for the old timer, it's the call to work a little harder than usual. "Code 99" – FGD (found on ground dead), tits up, stiff. In EMS, it means we are too late again, but if we try really hard, we might get a vegetable (derogatory term used for those who have the heart resuscitated, but not the brain - brain dead, you're too late again, but

24

at least you have a code save, another notch in your EMS belt.)

So, here we go again, "code 99." We walk up to the door loaded down with all the kits we need to resuscitate the dead. We knock. If the dead man opens the door, I'm out of here. No one answers. We look in the window and see a little old lady slowly moving, no, sliding a walker on the hardwood floor. A petite, wrinkled, stooped over woman who is older than dirt slowly turns the doorknob and the door creaks open. With a worried, careworn look, she states that her husband of sixty years (being married for sixty years is amazing to most of us) is missing. She informs us that early yesterday morning at around 6am, her husband went out back to do some gardening and he has not yet returned. It is presently 3 pm, mid-afternoon. She also states that Mr. Statue loves to go out and work in his garden for hours.

"I would go check on him myself, but I just can't manage the steps down to the back yard," she states in a forlorn tone.

We make our way through the 1920s antique furniture and the stacks of newspapers to the stairs leading to the backyard. At the foot of the stairs is an open carport with a sawhorse and a stack of firewood. Standing, no leaning on the sawhorse is Mr. Statue, holding a gardening trowel with a death grip. He had finished gardening and was on his way into the house when he had a massive heart attack. The sudden death froze him into place. Kind of like the leaning tower of Pisa and the Statue of Liberty combined, but dead.

I reach over to check his pulse, but he is as cold as a Popsicle. I am glad I don't have to go up and inform Mrs. Statue of her husband's demise. The firefighters are the lucky guys to get that job. It would have been very hard to explain that even though her husband is still standing, he is in fact dead, and she will have to get someone else to do the gardening.

Roll Over Dozer

Don't lose your head when your bulldozer starts to roll over.

Just bend over and kiss your ass goodbye.

How fast do you have to be going before you can roll a bull-dozer?

Crank the wheel really hard, at 20 mph in a car or jeep and you might roll it over. But a bulldozer can't even reach the maximum speed of 10 mph!

I know! Drive the bulldozer up an unstable hillside and close to the edge of a thirty-foot ravine.

Now you can roll the bulldozer.

Do this as you kiss your ass goodbye, and it will land on your head, this causes a splatter effect, kind of like popping a pimple.

Throw your arms and legs out spread-eagle and call the MX. Better yet, have someone else call the MX because you're going to be a little busy.

This is how we found our bulldozer diver.

Spread-eagled at the bottom of the ravine with his head under the bulldozer, or rather what is left of his head? He kind of reminds me of someone doing snow angels, except in gravel. He must have flailed around as his head popped like a pimple. Sorry, not a pretty picture.

As I stood at the top of the hill and looked down on Mr. Spread-eagle, another worker came up to me and asked me if I thought he was going to make it.

"No," I said, "I think he's dead."

Usually that's what happens when a bulldozer lands on your head, I thought.

I love stating or thinking the obvious. Kind of like blue comedy hour, 'Here's your sign.'

When the worker heard this, he turned a lovely shade of white and vomited at my feet, just missing my shiny boots.

To add insult to injury, the worker stated that the dozer driver was just married, and was not certified to drive the bulldozer. "He will probably lose his job," the worker said.

I don't now what being just married and not being certified to drive a bulldozer have in common, but he made a connection somehow. I don't think the demise of the dozer driver had quite sunk in.

The dead bulldozer drive won't be able to come to work anymore, at least until he is certified, let alone support his new wife.

When you're dead, do all those certificates really matter?

Talk about losing your head and all your cookies at the same time.

I guess the dozer driver will be getting a certificate though, a death certificate! Another one to hang next to that marriage certificate, and maybe they could get a bulldozer certificate to him posthumously. The certificates might look good on the wall next to his family portrait.

Does that make him certifiable? He was crazy to drive a bull-dozer over the side of a cliff while kissing his ass goodbye, anyway.

I now know what I learned from this young bulldozer driver, how to roll a bulldozer with style.

I know this is a brutal fatalism, but such a waste was hard to rationalize. Emotionally, or any other way, for that matter, this is one way we paramedics often cope with tragedy, especially a senseless tragedy like this one.

Mr. Bow Jangles

Another random call of a man supine, that is lying on his back. CODE 99! Here we go again. As we pull up to the house, we see that the firefighters have already arrived. The door to the house is open and we grab our gurney and scurry into the house. The firefighters have removed Mr. Bow Jangles from his old dilapidated bed in which, per his family, he was resting so quietly. He has no pulse and he is not breathing, another caring person called 911 for the sleeping dead. Haven't they heard of not disturbing the dead? Is there no rest for the wicked? Mr. Bow was dreaming of angels as he was rudely yanked from the sagging bed. The bed was imprinted with his bowed stiff body. A cameo of Mr. Bow at rest. The bed would not accommodate doing CPR, it was too soft, so the firefighters have placed him on the hard wood floor, a perfect surface for the job, unless of course, you have bad knees.

Frantically, they begin the resuscitation process: CPR, assist breathing by inflating his lungs with a bag valve mask, and push on his chest to compress the heart between the ribs and the spine, simulating a beating heart. Start an IV and pump drugs of adrenaline and atropine into the veins and hope for the best. Asystole – no electrical activity detected in the heart, fundamentally dead. This is what we find as we walk into the home of Mr. Bow.

As I walk up to offer my assistance, I notice that something is not quite right. Each time they compress on his chest with CPR, his feet shoot up into the air. Mr. Bow has taken the shape of his bed, a bow-shape. His resuscitation has become rocking chair CPR. Mr. Bow has been dead for several hours and rigor mortis has set in. I point this out to the frantic firefighters and the efforts to revive the dead are terminated. They will have to use a lot of padding in his coffin. Maybe he could be buried in a hammock. He would have to be laid to rest in his position of comfort. The firefighters learned the hard lesson of triaging the dead. Next time, they will check to see if their patient is rescuable.

Chapter II

The Bizarre

(not a three-ring circus)

This chapter is summed up in one word – bizarre. Strange, odd, and unbelievable would also be applicable, but bizarre has a circus flavor we like. Sometimes, that is what we feel like, that we live in, and are members of a three-ring circus. We have become experts in dealing with these odd and strange situations. So let's go to the Bizarre!

Mrs. Fly

How do you describe foul odor? They say that words are inadequate when you want to convey the smell of something. I thought about including a 'scratch n sniff' here, but I don't think they make scratch n sniffs for the kinds of odors we are going to be talking about here.

So scratch yourself; no, that won't do, not at all. '*We all think our own shit don't stink*' – an important fact to remember as we introduce you to our next patient.

We all know the smell of ammonia or that stinky fart from grandpa, so just imagine the smells we are going to try to describe in the following encounters and remember the diaper and grandpa's farts. Memory and smell are very strongly linked. We forget what we see, hear and touch, but not what we smell or taste. Recalling those smells, consider the following.

The door swings open and a firefighter runs out into the yard and spreads his cookies onto the lawn like fertilizer, that is he vomits in the grass. This is a pretty good sign that something is not right.

With apprehension, we walk up to the door and a tsunami of odor assaults our nasal passages.

Another pale firefighter pushes past us gulping for air like a fish out of water. He leans on the porch rails and tells us that this lady rolled out of bed and they can't get her up.

How does this smell come from that?

I look at my partner for some rationality, but he has the same confused look on his face.

I take one last breath of fresh air as I prepare to enter.

I go in!

The white sheets on the bed are stained with brown and yellow streaks, and the bed is littered with food bits, urine stain and feces.

My eyes water and as I look through the teary vision, I see a large naked lady lying on her back on the floor. A brave young firefighter is shoving a blanket underneath the woman trying to make her ready to move. His objective is to get her off the floor and away from the smell. I can tell this by the bluish-red expression on his face.

I go to his aid out of instinct (note the pun, I know bad pun).

As I go to turn the woman, the layers of skin that make up her abdomen separate and I see a nest of maggots that have made a home in this poor woman's fat. I now know where the odor came from. I try and hold my breath and do my job. What is absolutely amazing is that this woman is alive and complaining about the hard floor. She tells us she has been lying on the floor all day. She complains that her back hurts and that she is hungry. Maybe she needs some food for her nest of babies? The odor is bad and I hold my breath as best I can.

No one told me that I would have to practice anaerobic para-medicine.

My mind fills with the memory of firefighters puking on the yard moments ago, and I consider joining them.

No, I'm the few! The proud! And the brave! And the insanely dedicated!

I get the woman up and on to our gurney, carefully keeping the nest of maggots tucked away. We wrap her in a yellow blanket hoping to contain the little ones and the smell.

Our goal: get out of the house! Maybe fresh air will help.

Mrs. Fly was left along in bed with no one to help her, but a small child, who brought her sandwiches.

No one helped Mrs. Fly to the bathroom. No one helped Mrs. Fly change her bed, or shower.

This contributed to the smell of course, and the nest of maggots.

I guess you grow accustomed to your own stink. She didn't notice that foul odor, as far as I could tell.

I told the nurse at the ER that the patient had some friends in her abdomen. She looked at me like I was making fun of her for some odd reason. Soon, I saw her scurry from the room to the bathroom; she was almost as pale as… well, maggots. I think she found the younglings of Mrs. Fly.

Believe it or not, the maggots kept Mrs. Fly alive. The maggots ate the dead rotting flesh and saved Mrs. Fly from getting gangrene in her abdomen.

Maybe they were her babies? They were in her abdomen, anyway. Mrs. Fly did live and the baby flies, well who knows? I think the whole thing is just gross.

Mrs. Pendulous

Our next story also addresses bad odor, but this story adds another factor, in fact, several more factors.

We all know that shit stinks. The odor of rotten or spoiled meat is nasty. We know the rank odor of spoiled food. We are acquainted with the smell of the locker room body odors and the nasty stinky sock smell.

Some are even familiar with the pungent odor of the barnyard and the meat market.

Some can recall the smell of blood and the cat litter box that needs changing.

That brings me to the next call: combine all of the above odors and there you have her, Mrs. Pendulous.

This time, no one runs from the apartment because you're the lucky one.

You're the first! Tag, you're it! As you push open the door and see into the apartment, it is littered with newspapers and cats.

Yes – live, meowing, purring pussies.

Crapping on everything available; Dish towels, couch cushions, counter top, stovetop, floor, flat surfaces, and any newspapers to be found. Whatever happened to the adage "Don't crap where you sleep"? Where in the hell are all these cats sleeping? We have just walked into their litter box. To say there are about twenty cats in this small one bedroom apartment is probably an underestimation.

I am sure there are more pussies in this apartment than in the Broadway play *Cats*.

No, the cat odor here is "fowl" (sorry about the pun, but I couldn't resist) to say the least; but what was that other smell?

As I search around the apartment for our victim, avoiding the cat piss, crap and scurrying kitties when I can, I find, sitting on the bed, in a dark filthy bedroom is Mrs. Pendulous.

As my eyes grow accustomed to the light, I see a sight I cannot quite get my mind around, an old woman leaning over a trash container, spewing coffee grounds. (A bad sign.) Her blouse is open and her breasts are resting on each knee like bologna sausages covered in coffee grounds, yet all wrinkled and old. Between her breasts protrudes another sausage-like appendage, kind of like pepperoni yet larger. This is her umbilical hernia, so I was told. It too, is covered in coffee-colored emesis (vomit containing digested blood). This middle appendage dangles just above the cat piss and is being cuddled by an amorous kitty cat. As I hold back the nausea, I ask the old woman if we can move her onto my gurney. She says yes and then belches more coffee ground vomit.

When she is through with this episode of disgusting bodily function, she says, "Just don't let my cats get out."

Ah, the reason for all the crap! I thought. No kitty litter and trapped in this apartment from hell.

I call back to my partner to cover the gurney with a yellow blanket in order to protect it from the terrible smell that exudes from the old woman.

As Mrs. Pendulous stands, tarry black feces drips from between her legs. There is that other odor! I think, as I gag and try not to vomit. If you have ever smelled bloody stool, you would understand. We take Mrs. Pendulous to the hospital, but to no avail.

Mrs. Pendulous is going to have to get someone else to watch her cats.

Mrs. Pendulous is now making heavenly coffee. She has gone to that great litter box in the sky.

If you look real hard at that cloud formation, it kind of looks like three sausages suspended over a litter of kittens in a coffee ground litter box. People usually see fluffy bunnies and teddy bears in the clouds, but I see the latest insane call.

Welcome to my world. Watch where your step, there may be a pussy underfoot.

Mrs. Pool

Here we go again. Another 'Bleeding problem' comes over the radio. We respond code three to a rehabilitation facility, better known as a nursing home. Will it be a ground level fall with a skin tear, a gastrointestinal bleed, or some other unknown form of bleeding problem? Such a loose term like 'sick person' makes it hard to get mentally prepared for the true nature of the call.

I walk into the facility and am directed to the patient's room. Upon entering, I see a pool of blood covering the floor. The trail of blood leads us into the restroom. My gut tells me this isn't a GI bleed because it doesn't have that distinctive odor. I turn the corner to the bathroom and there is Mrs. Pool sitting on the toilet. Her caregiver has a large white, well, used-to-be white towel wrapped like a tourniquet around Mr. Pool's lower leg. Mrs. Pool had necrotizing fasciitis, also know as the infamous bacterial flesh-eating disease. This bacteria was eating away at her calf muscle and had to be surgically removed to prevent Mrs. Pool from becoming the main course.

While sitting on the toilet, Mrs. Pool bumped the back of her leg on the toilet. The floodgates broke open and the blood started to spill out.

I remove the towel to see what we have. She still had some kind of thin bandaging surrounding the wound. I look around the back side of her lower leg and there is a steady flow of blood coming out of the back of her leg. I take some bandaging and apply it to the wound with pressure. Mrs. Pool looks like she is going to pass out from loss of blood so we pick her up and put her on the stretcher.

Her leg is still hemorrhaging so I take my gloved hand and put it in the gaping hole and feel for the bleeding vessel, and then apply pressure. I feel a throbbing pulse and press harder with my fingers inside of her leg. It is a gross feeling to say the least. I keep her leg elevated and control the bleeding in this position all the way to the ER. By the time I get to release the pressure and get my hand back, the bleeding has stopped.

Time to wash up. My hands still feel dirty even though I was wearing gloves. I scrub several times that day, but still have the creepy crawlies thinking about how close to a flesh-eating bacteria I was.

Mrs. Zombie

We are called to a house fire and explosion with a victim with severe burns. As we arrive, firefighters are attacking a two-story house fire in which the flames are spewing from all the openings of the house. Every window and door is on fire. Two firefighters are escorting a woman out to the sidewalk. She has her arms out like a zombie and her face is swollen to the point she has lost sight. Her face is burnt from the explosion. Skin is dripping from both her arms and hands. Her screams grow louder and louder. She has no eyebrows. Her hair is crispy black and gray with ashes. The woman continues to scream as we lie her down. She is burnt from the waist up.

Mrs. Zombie decided she did not want to live anymore, so she turned on the gas on her stove and blew out the pilot light. Then Mrs. Zombie reclined on the couch and waited. She waited for the room to fill with gas, and then she lit up a cigarette.

After we load Mrs. Zombie into the helicopter fully immobilized and sedated, we go back to investigate the scene of the crime. When you walk into the room, from the waist up, the walls and ceiling were charred black. The cigarette was the fuse, the gas the fuel, and the resulting explosion blew out the back wall of the house. It ignited the curtains, thus the flames shooting out the windows.

Mrs. Zombie's burns were a result of the explosion, but the couch was her "airbag," preventing her from being blown out the back of the house.

Mrs. Zombie walked out like a zombie, stunned until the pain of the second and third degree burns hit her like an explosion of a gas stove. Go figure.

We had loaded Mrs. Zombie into the helicopter and she was taken to the burn center, where she will spend months recovering, having her dead skin scraped from her body and receiving multiple skin grafts.

She probably will never have eyebrows, and people will always stare. She will probably never smoke another cigarette without having flashbacks of this day.

Mr. Titanium

Sometimes, when you're young and full of stupidity, you just can't get enough EMS so you volunteer as a firefighter. EMS to the young is like heroin, only it's adrenaline, much more addictive. You become addicted until you realize it's killing you. Even then, you can't stop. You want more. You can't get enough. They should have rehabilitation centers for emergency personnel.

Yes, those were the days. This brings me to my next story.

Mr. Titanium was a "hoarder" – that is someone who can't bring himself to throw anything away. Some say it is from going through the depression. Some say it is a compulsive disorder resulting in keeping every piece of junk mail and newspaper delivered to your door. A fear of throwing away a part of oneself or something important, something essential to life.

I know that the root of the problem is laziness, but everyone is afraid to say it.

Their mother never made them take out the trash, thus they never take out the trash. There I go again, being insensitive and stating the obvious. It's a bad habit I will have to break some day.

The papers and garbage becomes a monument to their laziness. 'Lifestyle of the poor and lazy' reminds me of the antithesis of the TV show 'lifestyles of the rich and famous'.

Their mantra, "Don't touch my stuff" – little do they know, they are preparing for martyrdom.

Gathering the kindling around them so when the time comes, just the right spark will light them up like a roman candle. A celebration of their own personal crematorium. So, Mr. Titanium was preparing for his lighting ceremony. But Mr. Titanium was preparing for his lighting ceremony with an added special twist. Mr. Titanium was always cold so he had set up a system of propane heaters strategically placed throughout the house. Each of these heaters was connected to the propane tanks by rubber tubing –

flammable rubber tubing. The scene is set for his glorious exit and his introduction to the gates of Hades and the fires of hell. Just the right spark ignites the trash and melts the rubber tubing. Each propane tank becomes a blowtorch, like a furnace in a crematorium.

When we open the door of the smoking furnace (the house of crematoria), the flames belch from the door like a dragon from hell.

The flames are so intense; we cannot remain on the porch, but are driven back into the street. I swear I see glowing eyes and a flaming tail, but no, that is the propane tanks spinning as they discharge their contents onto the pillars of newspapers. The newspapers look like the scales of a dragon as they ignite and float to the ceiling, being driven by the super heated air. Within the belly of this dragon sits Mr. Titanium, seated comfortably on his recliner chair. He is awaiting the introduction to the god of laziness – Lucifer himself.

Once the propane tanks have burnt themselves out and the dragon is quenched with gallons of dragon repellent (aka water), we begin the arduous task of digging through the charred remains of the home of Mr. Titanium.

As we sift through the ashes, which are pretty much all that remain, our shovels strike something shiny and metallic. It is the titanium knee replacement of our victim. As we dig around his remains, we uncover the 'recliner chair' the only remaining monument to his demise, holding his charred remains and a titanium knee. A symbol to us all not to be lazy! This is a real

41

motivation to take out the trash, unless of course you wish to create your own private crematorium for your cremation.

Mrs. Bird

Sometimes, our system of social workers and police officers come together and they dial 911.

What I am talking about is the infamous 'welfare check'. This is the scenario when an individual, who lives alone and usually rejects the help of others, needs help, but they don't call 911. Often, they are reclusive and love it that way. The hermits amongst us – that elderly lady who has been labeled the local witch or scary lady down the street by all the school children. The kids cross the street out of fear, instead of walking too close to the dilapidated old home. Fearful of the witch's curse and the stories that abound, their imagination has persuaded them the stories are true.

Mrs. Bird is one of these ladies. Her daughter usually checks on her weekly, but she lives out of town and her only contact is by phone. Last week, no one answered the phone.

Sometimes, Mrs. Bird is that way, but the daughter is concerned this time because Mrs. Bird has been in poor health. So the daughter called the police and the social worker to check up on her.

The lawn is overgrown and the house is in need of repair. The door is locked and the smell exuding from the home is one of death and decay. The police and the social worker call us before they enter the home, knowing something is not right. When we arrive, the police have opened the door and are using flashlights to see within this tomb. It is in the middle of the afternoon and the house is 'pitch black'. The windows are covered with newspaper and duct tape. Each door is decorated with spider webs like in Halloween, only real spider webs. The living room is a labyrinth of newspapers and trash.

The odor is one I recognize from when I was a child. I loved to bring home things I found and one of those things was a dead Robin. The eyes were rotted out and maggots infested it, but I thought it interesting. My mother did not. That was the smell.

We announce ourselves into the darkness and strangely enough, we hear a weak cry coming from a back room. We make our way back to the room from which the cries came and find Mrs. Bird. She is sitting on the toilet and she tells us she can't get up. Body odor is now the predominant smell and that mixed with that dead bird smell causes one of the officers to flee the scene. I think he went out to check the yard for possible evidence of a break-in; that was his excuse, anyway. Mrs. Bird has been sitting on the toilet for three days. She tells us that she can't feel her legs and that she doesn't want to go to the hospital, but wants help to get up and put on some pants. As we help her up and hold our breath, we notice the room across from the bathroom. Our eyes have grown accustomed to the dim light and we see the reason for the bird smell. In this room are about ten birdcages. In each cage are several birds that have suffered the agonizing death of starvation and dehydration. Some hang from their perches and some lie at the bottom of their prison, but all dead. Mrs. Bird seems to have no remorse for the demise of her birds.

We persuade her to go to the hospital to be evaluated, the social worker gets involved and states she will meet us at the hospital to make the situation known to the powers that be. We also find out that the electricity had been turned off for three days and that the place was to be condemned. Mrs. Bird lost her birds and her home. She almost ended up like her birds; trapped and dead, all because of her hermit life style. It was baffling to us that this woman chose to live like this.

Mr. Spoon

When the plumbing stops, it is absolutely amazing what people will do to get it going again.

What I am referring to is the problem of constipation. This has become a very serious problem amongst those who take pain medications, and those who find their bowels responding to the ravages of age. Many have turned to laxatives and enemas to address this gut-wrenching issue.

The joy of a bowel movement has evaded them. They have entered the world of hemorrhoids, prolapsed rectums and bowel impactions. The medicine that has delivered them from pain has betrayed them to the dark world of constipation and impaction.

This brings us to Mr. Spoon. We are called to a patient with

44

abdominal pain, also known amongst the more lighthearted medics as *abominable pain.* Code three, here we go again, the patient is having an emergency stomachache.

We are met at the door by a young woman, who states her father is in the bedroom, requesting transport to the hospital. As we make our way through the quaint farmhouse, we find an elderly man lying on his bed under the typical bed linen. His demeanor is very matter-of-fact with a tinge of orneriness. His complaint is that when he moves, his stomach hurts. As we interview him, we find that he hasn't had a bowel movement for some time. We prepare the gurney to move him and as we pull back the linen, we notice bloodstains on the sheets. My partner, an EMT-Basic, looks at me with concern and I ask Mr. Spoon if he has been bleeding for very long. With that same matter-of-fact attitude, he reaches over to a shelf and picks up a ladle-sized spoon and holds it in my face. The odor of feces is noticeable as I pull back. He states, "I have been using this to try and dig out the crap from my ass."

I notice the brown and red stains on the spoon and the expression of disbelief on my partner's face and with as much resolve as I can muster, I keep a straight face. The sheer mechanics of what Mr. Spoon has done was very difficult enough to imagine. To use a large spoon to cram up your ass just to get the poop out was an act of desperation to say the least. The sheer mechanics of the act was unfathomable.

I recently had to get a colonoscopy and awoke on the table as

45

the doctor was cramming the scope up my butt. I can tell you it was not very comfortable. To think, this man was doing a similar procedure on himself without being medicated gave me a whole new appreciation for Versed. What is sad, I could almost relate – GROSS! Of course, the tube wasn't near the size of his spoon.

We move Mr. Spoon to our gurney and take him to the hospital of his choice.

Mr. Spoon has gone to that great ladle in the sky. He acquired a severe infection from his endeavors, from which he never recovered.

Lesson: Spooning the ass can kill you, especially with a ladle-sized spoon.

Chapter III

That Just Ain't Right!

Obesity is an epidemic in America! There, someone not selling a weight loss product said it.

This epidemic sets EMTs up for injury and hemorrhoids.

We sit in our cramped ambulance and then, without notice, are required to lift an ungodly amount of weight. No wonder we develop back pain and other debilitating injuries. When we mention obesity, we are not talking about the individuals who might need to lose 50 pounds. We are referring to the person that is so overweight they are literally confined to their homes. Some can only move just enough to get into the kitchen for a snack, or meal, and then onto the bathroom to complete the cycle, no not bicycle, which is what they really need. The habit of over-eating has become an obsession. These people are in desperate need of a weight loss plan. It's sad and often disgusting. Consumption of too much food and a sedentary lifestyle has jailed these people in their own bodies. The stories that unfold take you into our reality of the morbidly obese and their dilemma of sustaining a level of normalcy. These people are walking (usually sitting, but it doesn't quiet fit) time bombs, it's just a matter of time before they explode, and make that call for help. In our relating of these stories of the extremely obese, we make observations and comments that may be construed as making 'fun of' or insulting, but that is not our intent. We are confronted with the dilemma of the morbidly obese, and this confrontation is a challenge, both mentally and physically and psychologically for us all. Those of you who are paramedics will hopefully relate to the issues that arise in our stories and we hope all those who read them will understand our struggle to maintain an attitude of compassion in the face of the daunting challenges set before us when we deal with such large patients.

Mrs. Caterpillar

We receive a call for a lady who has fallen and can't get up. This *I have fallen and can't get up* problem has become an epidemic. It is amazing how often 911 is called for this epidemic.

If only I could invent an antigravity belt for the vertically challenged / horizontally impaired. Or maybe some fizzy lifting drink like the kind Willie Wonka had in his amazing chocolate factory. That image that pops in my mind is the one of Charlie and his grandfather drinking the fizzy lifting drink and floating into the air. Of course, this would mean we would have a lot of floating dead people, but some of them would be alive and they would be spared that broken hip or fractured bone. We would need antigravity deactivation devises to get our patients off of the ceiling (or just get them to burp) – another reason to call 911, but what the hell, another EMS challenge. I digress, back to our story.

As we walk into the mobile home, one of those 70s homes with metal walls and thin uninsulated windows, we hear a woman calling for help. This metal box someone has called a home is somehow familiar. Yes, I know I work in a metal box they call an "**ecnalub-mA**." That is what is on the front hood, anyway… maybe they should put "**emoH eliboM**." No, that doesn't work. Well, it was a thought.

As I walk into this so-called home, a recliner chair and television fill the living room, which is part of the kitchen, both almost indistinguishable due to the clutter.

Food is littered like a Hansel and Gretel trail from the kitchen to the hall. I'll just follow the trail. Fortunately, the trail has not been eaten by the wildlife.

Maybe I'll find the 'fat' lady. You know the job's not done until the fat lady sings. Yes, there she is at the end of the narrow hallway trapped in the doorway of her bedroom – face down, squiggling like a caterpillar. This is what it looks like from my vantage point, anyway. She is trying to get up. Of course, all you see are the rolls of flesh that make up her legs – spasming with the effort of each attempt to right herself. Kind of like one of those ugly hairless caterpillars on *Animal Planet*. The ones that can eat whole trees one leaf at a time. I'm sure she found herself in such a state one candy bar at a time. Now the challenge – to get the trapped Mrs. Caterpillar up off of the floor. This is something that I have observed about the nature of fat. It is has constant shifting center point of gravity. This is a very difficult problem, especially when you're trying to get a caterpillar off the floor. Each attempt at lifting a person with shifting fat seems to be an effort in futility. When we lift at the legs, the fat shifts to the abdomen or torso and then throws you off balance. It's like a shifting bowl of jelly. Imagine walking across a room with a bowl of jelly except without the bowl. The bowl is instead a plastic bag. You can't drop it or let it hit the floor. This is our challenge with Mrs. Caterpillar.

Another aspect of our dilemma was the hallway. It was so narrow, only two could fit on each side. This meant about 200 pounds of shifting jelly per assistant. The solution – contain the jelly! To get Mrs. Caterpillar up, we place a sheet around her center and get as many hands on the sheet as physically possible. Presto, a standing caterpillar. The key here is to not let her fall again. You see, she has knee problems. They can't support the jelly anymore so they lock up. They go on strike. The knee union has declared a walk out. When this happens, Mrs. Caterpillar finds herself on the floor again. Short-term solution: get a chair and have her sit. Long-term solution: Jenny Craig.

Let me tell you about the condition of Mrs. Caterpillar. Her legs have become so large from fat build up that they have lost their circulation.

This leads to all kinds of wonderfully disgusting diseases: Cellulitus, phlebitis, DVT's, and skin conditions to many to mention. These conditions often lead to amputation, not giving her a leg to stand on, and with her girth, this is a real problem.

Mrs. Caterpillar has finished her last candy bar. The cupboards are bare. Maybe if she spun a cocoon and metamorphosed into a beautiful butterfly, she could be free, but I doubt it. For some reason, I just can't get the picture of Mrs. Caterpillar squirming on the hallway floor out of my dreams … *what dreams may come?*

(Sometimes I think Hamlet and I have a lot in common, this is not good considering he died at the end of the play)

Mrs. Tsunami

"Code 99." A patient receiving dialysis has stopped breathing. The nurses know how to connect and monitor complicated dialysis machines, but when the patient dies, they go into a panic and call the lowly paramedic. 911 is their answer, we are their saviors. This concept makes us very conflicted because we do not consider ourselves saviors, just someone wanting to make a difference and lend a helping hand.

On our arrival, the victim has been moved into a separate room. A ten by ten room, which is writhing like a pail of snakes – full of rescuers and one patient. Mrs. Tsunami is over 500 pounds. She is five foot four in every direction. The room is of full to capacity. "Assholes and elbows," is a term often used to describe such pandemonium. My job – move Mrs. Tsunami to the ambulance, and at the same time, continue CPR and the drug therapy used to resuscitate the dead. The challenge: move 500 pounds of shifting, rolling fat onto a flat board. With this accomplished, another problem arises. The board is too small. The board disappears under Mrs. Tsunami like the beach under a Tsunami. Solution: add another board. Problem: two boards won't fit on our gurney very well, but somehow we manage.

Now the balancing act as we move her to our ambulance.

Mrs. Tsunami arrives at the E.R. with CPR in progress and all the medications we can throw at her, there is still no pulse or respirations, but we are a tenacious bunch and we will go the distance. With the help of the fire department and E.R. staff, we unload her from our ambulance and move her into the E.R. As we move her to the E.R. stretcher, a shift occurs from somewhere inside Mrs. Tsunami. Her fat becomes a wave of jelly. Everyone moves as one to prevent her from going to the floor, and when it seems all is well, the second shift of jello begins, like a Tsunami. All hands have moved to one side and little old me on the other side sees a huge wave of Mrs. Tsunami coming right at me. Little me, all alone, about to drown in a tidal wave of fat. My life flashes before me and I see the headline news. 'Medic succumbs to a wave of fat and is crushed by an obese woman in E.R.!' Much to my relief, the wave comes to rest on the stretcher, the center point of gravity. The news media will have to go somewhere else for its sensationalism. Mrs. Tsunami was declared dead by the doctors. She had finally succumbed to the tidal wave of fat. I have been spared the horrible experience of death by fat. Another memory that I hope will be lost to my old age dementia.

Mrs. McDonald

There are times when "911" is not the nature of what we do. Sometimes, we are virtual cab drivers, without the tips and ticker of course. We get paid bi-monthly, not for each time we have a customer in our ambulance. Today, we have a woman who is so large she is unable to fit in a wheelchair. Even a heavy-duty wide width wheelchair will not meet Mrs. McDonald's needs. Today, I wish we would get paid by the pound.

She will need the superhuman strength of the ambulance driver. Mrs. McDonald is over 500 pounds.

She has obtained this impressive girth by a very disciplined lifestyle.

First, she has not done any exercise for at least 2 years. When I mean exercise, I mean walking, sitting or standing. She has exercised her elbow when she lifted the fork of fat laden food to her mouth, but that was only because it was necessary. The key to such discipline is dedication and sheer strength of will; of course she has also had family support. Someone had to go to the store to buy the food. Someone had to prepare the food. Someone had to bring the food and someone had to feed the food to her. Otherwise, she would have done too much exercise.

Mrs. McDonald did something today she has not done in two years. She stood up. It was a gigantic effort, mind you. We had wheeled her into the home on the gurney. We did not raise the gurney because we were fearful that it might collapse under the strain. So we left the gurney down. This facilitated her so that when she moved her legs over the edge, she would be able to stand up.

As she stood, I find myself standing behind the largest pair of butt cheeks I have ever seen. What is frightening is that they are gyrating towards me as she is attempting to move so she can sit on her bed. I also find myself between her and the bed and my life flashes before me again. I can see the headlines: "Paramedic Smothered by Butt Cheeks."

(I hope you catch the pattern here. We paramedics have an aversion to media and the headline news. I think it is because we feel the news wants to candy coat us or vilify us. We are not sure which and this make us a little nervous.)

Thank God for quick legs. I am able to maneuver around and get out of the way of Mrs. McDonald as she sat on the bed. She is quick to order lunch after all that exercise: ten Big Mac burgers. Believe it or not!

Someone was actually going to bring them to her, too. They must have had a good life insurance policy on her.

I think they were trying to kill her. I think they wanted her to her eat herself to death. I guess it wouldn't be murder? So they would be able to collect the insurance. No, they forgot about all the hospital and ambulance bills. I guess they could always fatten up grandpa for the kill.

Mr. Tree

He is six foot seven, 650 pounds. His ankles are the size of small trees. In fact, the skin on each ankle is kind of like bark. His heart cannot keep up with the demands of his body. His lungs are under the pressure of the massive weight gain around his abdomen, and each breath is a chore.

If a tree falls in the forest, does someone call 911?

Mr. Tree fell in the kitchen and the entire neighborhood heard him, but no one called 911.

What the hell was that? An earthquake, or a car accident? Hope someone calls 911.

His wife, not much smaller than him, came home. She called 911.

Assholes and elbows, again as we walk into the home of Mr. Tree.

Firefighters are scrambling to rescue Mr. Tree.

Someone save a tree. Someone hug a tree. Someone do something!

He had stopped breathing when they arrived; all the weight in his abdomen finally crushed his lungs. They started the process of resuscitation. A tube was placed in his trachea, a feat in and of itself.

Mr. Tree's head is so heavy, it takes two firefighters to lift his jaw up enough to place the tube in. Compressions are a gargantuan effort. A firefighter is pushing on his chest, no jumping on his chest, trying to move his chest down just enough to get a pulse. He is jumping into the air and landing on his chest with all his weight.

Now for the gross part.

Mr. Tree has a baby tree... well, it kind of looks like a baby tree.

He has an umbilical hernia. It looks like a small tree stump protruding from his belly button.

It too, is covered in bark. It kind of freaks out the fire fighter doing CPR because each time he pushes, or rather jumps on Mr. Tree's chest, the tree appears to grow. This tree is coming out of his abdomen, kind of like in the movie 'Aliens', only with bark on it. It is nothing like any of us has seen before and we can't help but stare at this bizarre phenomenon. With all this effort, Mr. Tree starts to breathe and his heart begins to beat again. His lungs have filled with fluid and he is drowning in his own fluids, but at this point he is treading water. 650 pounds of "floating fat man." I recall the movie *Dune* and am reminded of the Baron Harkonnen and his levitation device. I wish we had one of those. Mr. Tree was going to be near impossible to move into our ambulance.

We need to move 650 pounds of dead weight onto a gurney a

third the size of his width. Using blankets and straps and about a dozen firefighters, we accomplish the impossible. The baby tree continues to grow as we treat his condition. As we arrive at the ER, Mr. Tree dies for the last time. I wonder if Mr. Tree's baby tree can be saved? No, God save the trees.

Mr. Bucket

What the hell is going on? My senses, especially my olfactory centers (the sense of smell) are being assaulted by a god-awful smell. The pungent smell of a sewer comes to mind. We have been called to a man down. Inside a singlewide mobile home that has not seen mobility in quite some time is Mr. Bucket. He has been supine on the floor for 2 days. He was trying to go to the bathroom when he fell. He fell in the kitchen. Yes, the kitchen. He was too large to fit down the tiny hallway leading to the bathroom, so he just went where he could. The sink: what a great urinal. A small garbage can: what a great toilet. There were various containers and receptacles sitting on counter tops, lying on the floor, and sitting on the dining room table. Each one contained Mr. Bucket's excrement. He just shit in anything he could find and flushed it down the sink. It makes me sick to think of what he used the garbage disposal for. We just got out of there ASAP.

Mr. Bucket is a nice guy; he just can't take care of himself. Diabetes had taken half his feet and he just couldn't support all the weight he was carrying. I have another medic in the back with me on this one. I was attempting an IV, when the other medic was trying to dig Mr. Bucket's wallet out of his rear pocket. I am sure glad I was starting the IV. I don't want to stick my hand/forearm underneath Mr. Bucket's pants. The guy has buckets of shit stored in his home. My partner has a sick look to him. His face is filled with disgust and nausea as he pulls out a thick handful of pasty feces. This is Mr. Bucket's wallet. Mr. Bucket's reaches for it without a care and pulls out his ID and says, "There you go."

I think, sucks to be you my friend. My hands are cleaner. Can I offer you a new pair of gloves and a shower? I can't wait to get back to the fire station so I can rid myself of this filth. I haven't touched much of anything, but I just feel like I took a nose dive into a sea of human excrement.

Upon transferring care to an ER nurse, it was our time to go, our time to recover from the scene that was before us. My partner and I decontaminate our office and I find Mr. Bucket's ID card. It is smeared with fecal matter and I have to take it back to his room. Upon approaching the room, an ER Technician comes out tying on a full frontal gown to shield himself from this mess. His face is green and his hands pale. He looks up at me and I state, "You should have smelled his house," as I hand him Mr. Bucket's ID.

Mrs. Brown

It's the dead of night. Maybe I shouldn't say things like that, but it's dead out here. Nothing is going on. All we get is the occasional car driving past us as we stare into the night, out of our metal box. We haven't run a call in hours and I feel dead. No, I just want to sleep, but we aren't supposed to, wouldn't want to miss that next emergency. Maybe I'd wake up if I had some stimulation, like a real emergency.

Dispatch blares over the radio. It's us, our emergency. Code three to a fall patient. We arrive to an apartment complex. I enter to find a 375 pound kid on a bed. Where he doesn't cover the bed with his mass, there lay crumbs of food. He states to me, "She's in the bathroom." I walk through the living room by the adjacent kitchen. Lining the counter tops are baking trays, one after the other, each with its various amounts of crumbs and broken French fries. I take it with a grain of salt and head down the hall towards the bathroom.

As I turn to find my patient, I am greeted with a 600 pound ass staring me in the face. Not only that, but a giant rectum covered in feces. She covers the entire floor, much like her son in his bed, only this time it's in a tiny bathroom. My patient slipped off the toilet while defecating. She has taken a bath in her own feces. She is face down with her head propped back by the bathtub. She is complaining of lower back pain. How the hell are we going to get her out of here? I didn't take any engineering or construction classes that would have prepared me for this. I would like to take out a wall, but that's not an option. She is almost as wide as the space between the tub and the cabinet, and I have to wiggle my boot between her and the wall just to stand in the bathroom. Stepping in dog crap is one thing, but this is disgustingly gross. Fight the thoughts and smells that are assaulting my senses and focus on the task at hand. We have to get her out.

My partner and I manage to get a heavyweight blanket under the front of her chest and I tell her to turn towards my partner. We pull and she turns. Fire is attempting to take the door off its hinges and also help move her legs. When she gets turned from face down to on her right side, I hear a blast, an explosion of methane and fecal matter. A new, stronger smell invades my olfactory senses, methane gas; sure hope no one lights a match. Something just hit the wall behind me. This lady is a GI cannon. By the splatter effect, I'd say the projectile of fecal matter traveled at 500 feet per second. Hell, now I have to worry about the walls as I balance myself in this bathroom from hell. We somehow manage to get her on her back and get ourselves out of the line of fire. We slide her with our blanket mover and with all available hands and lift her onto the stretcher; we will see our chiropractors tomorrow. As I push the stretcher towards our metal box, I fertilize the lawn with my boots. I don't want to track this crap into my ambulance.

So what is the real problem? Yes, she has back pain (so do I now), but what about the ten or so empty baking trays on the kitchen counter, once containing some sort of processed frozen food? Did they just leave the trays sitting there or did they just have another daily ritual of gluttony? What about being 600 pounds? She didn't get this way overnight. The choices we make today will affect us later. She is killing herself with this sedentary lifestyle. She is a food processor and nothing else. I forget – she is a mom, but she needs to stop taking care of her kid. He is being turned into a food processor, too. He, too, will be so large he can't fit on the toilet. He, too, will slip off the pot and get stuck on the floor, unable to move. The cycle will continue and bigger toilets will need to be made to accommodate the morbidly obese.

Just one question – *where's the toilet paper?*

Chapter IV

Ode to the O.D.

We are all seeking happiness. In fact, it is a constitutional obligation to pursue happiness. The pursuit of happiness in an inalienable right. Of course, I think our forefathers probably didn't have in mind what we do, but we have a right to happiness. Some find happiness in what they do – a fulfilling career. Some look for happiness in recreation – the slopes of mountains or the thrill of the hunt.

One of the other most destructive seekers for happiness is the substance abuser, the addict. He is defined as the habitual user. He is utterly, psychologically, and physically dependant on his choice of poison, his idea of happiness. These are the stories of people looking for happiness in all the wrong places.

Probably the most destructive drug our society has had to face in this generation is methamphetamine. Also known as crank, speed, tweak, meth, crystal, ice. Whatever you want to call it. The names all have some descriptive attribute of the nature of methamphetamine. The drug increases the metabolism and release of adrenaline to give the feeling of euphoria and hyperawareness.

The addicts of meth become so psychologically dependent on the drug that it changes their personality. Paranoid and sociopathic behaviors become the norm. Tweakers are sometimes not even recognized by their own families after they have used for an extended period. Part of this is a result of the destructive lifestyle, part the destructive nature of the drug.

Hygiene and self-care become non-existent. Many develop sores that they pick at constantly. The tweaker ages at a rapid rate and because they do not sleep, they have all the issues of sleep depravation. Such as hallucinations, disorientation, memory loss and a sharp decrease in cognitive skills. One plus one no longer equals two.

The question for the tweaker is always, "Where am I going to find the funds to buy my next fix?" Most often, the answer is to steal. Even begging or borrowing is not out of the question. Of course, consequences become subordinate to getting "cranked." Life, death, eating, sleeping, bathing: all only optional. This means the concerns of others are not a consideration. People, even family are expendable – sold for the next shot of speed. Meth destroys the lives of all it touches. The following poem is a window into the world of the drug addict, particularly the "meth head."

The ODe

I wished to be high

I would try not to cry

O' to float in the sky

Should I spit in your eye?

I drive the porcelain bus

To the circle of rust

O' powerful god of vomit

To get to the summit

The high evades me

I think I need to pee

O' to that burning in my groin

I think I got it from that bitch's loin

To get to be high

I think I would die

Call 911! He is not breathing, he is vomiting

He finally spit in God's eye

He really did die

By an ADDICT

Mr. Spiderman

We hear the radio calling our name and the adrenaline courses through our veins.

Our hearts and minds kick into overdrive as we go to 'the overdose'. The 911 dispatcher states police are on the scene of a man who was found running around his house. "Ok? What could this be?"

As we pull up to the house, we see six firefighters and four police officers wrestling with a naked man. Our thoughts go back to the Greek Olympics. So we stand back and cheer, not willing to be participants in this particular event. "Watch out for that crotch hold. It's a killer," we yell. Wrestling a naked man is not our idea of fun. The firefighters and police soon discover the lack of places to grab, thus the success of Greek wrestlers.

We grab our gurney and yellow blanket to prepare for our human burrito, the best way to contain a naked man with so-called spiders all over him. The man has red welts all over his body that are self-inflicted. With a fearful gaze, he becomes more and more agitated. He begins to swat at his arms and head like he is trying to swat at a swarm of flies. With each swat, a red welt is evident of his desperate battle with this unknown enemy. His fear takes hold and he begins to squirm like a man covered in insects. He starts yelling, "Get them off! Kill the spiders, they're everywhere," he says with a look of terror on his face. "They are sucking out my blood."

With a desperate plea he looks at me and says, "Please! Get

them off of me."

The firefighters keep trying to reason with the man, telling him there are no spiders, to no avail. I tell the man I have the solution to his problem and I run back to the ambulance. I grab the disinfectant spray. I tell him this is our "heavy duty, all industrial spider and insect killer." As I spray the imaginary spiders, I tell him that it kills on contact. The man calms down. The firefighters look at me like I am either a genius or a certified member of the insane asylum; I'm not quite sure which. I look at them and whisper, "What the hell! It works." We lead the calmed spiderman into our ambulance and are told by the police officer that they found a cornucopia of drugs and paraphernalia in Spiderman's house.

Methamphetamine was Spiderman's spider attractor. I guess the naked running around the outside of the house was his attempt to rid himself of the blood-sucking spiders. Spiderman had been awake for three days straight. Sleep deprivation is mind-bending. The methamphetamine was his *No Doz* with the side effects of paranoia, hallucinations and self-inflicted welts. Spiderman took our disinfectant with him as we dropped him off in the ER. The nurse tried to take it away from him, but the spiders must have been all over her because he pointed and sprayed. The RN left the room with a mouthful of "industrial strength spider killer"; she, too, is now disinfected. Hope she isn't a tarantula. She looked a little hairy. Sometimes, we have to be creative to meet the needs of our patient and prevent wrestling with spider-infested naked men.

Mrs. Cigarette

"Code 1 and stage" – this means we are going to a potentially dangerous call. The police arrive at a nice home where a caring elderly gentleman greets them with a smile. Sitting on his porch is a middle-aged woman. He informs the police he found her sitting in his front yard shaking like a leaf in the wind. The woman looks up at the police and the officer immediately calls us into the scene. We pull up as the police escorts our patient into the ambulance. Mrs. Cigarette appears very anxious and acts paranoid, with her eyes darting about looking for imminent threats. What is remarkable about Mrs. Cigarette is that she has a disconnective disorder as well as being bi-polar.

Add methamphetamine to the mix and it becomes a bad combination. She has cigarette burn marks up and down both arms, and on her chest, neck and in a circular shape on her face. Each burn is at a different stage of healing... infection, festering, scarring. What makes this scene unsafe is that this woman is the breeding ground for every virus and bacteria known to man. And I get to put her in my ambulance. My metal box with a small vent fan, a home with a sign saying "Welcome!" to all the viruses. I think I am getting paranoid and developing OCD. Mrs. Cigarette was photographed by the police. I think they were making a documentary on what drug abuse will do to you. She was taken to a drug and psychiatric rehab center, after being treated with antibiotics and psychiatric medications.

Mrs. Cigarette will always have the reflection in the mirror glaring at her for the lifestyle she has chosen. I wish I could better describe that reflection; it wasn't pretty. All I can say is ***Don't Do Drugs!*** Go to your doctor for a prescription. He will give you the drugs you need. Maybe one day they will come up with a pill to solve all of society's problems. Mrs. Cigarette's burns will scar her for the rest of her life, and remind her of her pursuit of happiness.

Meth Madness

A young teenage addict is looking for items he can steal so that he can sell or "hock." He knows that the best place to get these goods is the suburbs. Each house is equipped with electronics, and hot electronics bring top dollar. These homes are not protected by any security devices and they are prime targets for his purposes. He walks up and down the street looking for a home with the lights out and no cars in the driveway. He needs a quick buck so he can score some more "crank." That rush of speed coursing through his veins is gone and the withdrawal of dependence is assaulting his senses. His propriety has abandoned him. He will kill for his drugs. He finds the back door unlocked and slips in. The man is taking a shower. Normally, this would deter the addict, but not tonight. He finds a VCR, a TV and a video game. These will do. Gathering these items up, the occupant hears the commotion. He sees a stick is in the corner of the room and he grabs it ready to kill. The man comes out and the addict begins his assault with extreme prejudice. His first blow is disorienting to the man. He stumbles around the house and the addict continues his assault! The walls become a canvas of the addict's making. Each blow paints the walls with spurting blood. Each blow is a desperate act to get what he wants. All sense of restraint has abandoned the addict and he beats the man continually. Each blow sprays blood on the ceiling, the walls and the furniture. To escape the vicious assault, the victim seeks refuge in nearly every corner and nook of his home. A trail of blood is left behind as evidence of his desperate search.

The man comes to rest under the kitchen table. Curled up in the fetal position holding his wounds. He is unrecognizable. Blood has soaked his naked body. His violent death is covered by the cries of another. The man has a child. The addict thinks in his warped mentality that these cries must also be silenced. The addict finds the child in bed calling for his father. He takes the stick and beats the child into unconsciousness. He grabs his stuff and leaves. His act of murder has fallen on deaf ears and he escapes to his next fix. It is late at night when he escapes.

A friend has come to see why he has not come to work and the horror of the sight makes his hands shake as he dials 911. The police call us in. The child has awoken and moved into the living room looking for his father. A female firefighter arrives first and takes the child in her arms, moving him outside. I walk up to the officer and he states, "This is a crime scene," with a look of distress and disgust.

I ask, "Is there someone else in the home'" He shakes his head yes and tells me there is a body underneath the kitchen table. "Is the man dead?" I ask him. He is so stunned by what he has seen, he is unable to answer. I tell him I will go in and check and I will take care to not disturb any evidence. As I enter, I see the artist's work; a life struggle painted in blood. Some abstract work saying, "I have something to live for." I see the work of death; the destructive force of an uncontrolled desperate addict. As I come to the side of the man who has given his life for such a senseless cause. I find the epitome of the addict's work. The man is curled up, clothed in the blood of a hero. His own blood, given so that his son might live. The juxtaposition of the addict who took his life: selflessness vs. selfishness. We are able to life-flight the child to the hospital, and later we are told he will live. We are saddened by the death of the father and overjoyed by the news of the child's survival. A juxtaposition for every EMT: life and death. We are on the scene for three hours aiding the officers in gathering evidence needed to capture this *Freddy Krueger* of the meth world, trying to reconcile in our minds the purpose of such mindless mayhem.

When he is caught, the face does not match our idea of one who would commit such a crime.

This is what meth does. It not only changes your face, it changes your life. Life in prison for a seventeen-year-old is a long time.

Mrs. Southern Comfort

Apartment complexes are a pain in the butt! *Especially* for paramedics looking for a 911 caller, the labyrinths from hell are another reason we don't make it to calls on time.

No matter how many of these complexes you go to, you usually get the same maze. We respond to these places and are unable to locate where to go. It's very frustrating. I was supposed to be off shift about ten minutes ago, but Mrs. Southern Comfort had to scare her neighbor by hiding half under a parked car, crying her eyes out. She doesn't respond to my questions, her only form of speech a slurred outcry of "Daddy!" Okay...

Sitting her up, I hear an up-chuck sound and out pours a thick flow of half-digested pepperoni pizza and beer. It's a smell that makes me nauseous too, not the best smell in the world. We deal with all kinds of unpleasant odors at various times. Dealing with drunks and combative patients is always a challenge, especially when your thoughts are somewhere else. It's the middle of the night, it's cold, raining and it's cutting into my short weekend. These are the thoughts that go through my mind as we move her to our stretcher. Things change when the woman starts to throw punches at us. She hits my partner with a slurred punch in the jaw, and we put belts around her arms for temporary restraint. He blows it off as nothing. I wish I could just lace right into her right now, but I have a glass jaw. I'd never hit a woman, but she was really tempting. I wish I were a cop at times like this. I could just tackle her like a linebacker, but then again, I'd be wearing a nice coat of puke with a nice decoration of pepperoni here and there. On second thoughts, I do love hearing people when they get tazed.

Upon transferring Mrs. Southern Comfort to our ambulance, she frees her arms and starts throwing punches again. We tie her arms down to protect us from the abuse. Rage sets in, but I have to control this, it's my job and not my place.

Now she speaks with her slurred Polish accent, "Fuck you, asshole!" Finally, she speaks! She repeats over and over, a verbal assault instead of a physical one. I tell her how it's going to be. We are the nice way to go and the cops are the bad way. After my partner helps me restrain Mrs. Southern Comfort, he steps out to talk to the cops. He wants to see how they want to handle the situation before we start transporting. The verbal assaults continue. As I am writing information down, she flips out. It's like a switch turned on. Her legs come loose and rise into the air, kicking around. She turns towards me as I try to avoid her, but it's no use. I felt like a ship in a cheesy sea monster horror flick. Her legs have a hold around my waist. This is the "Kama Sutra" from hell. If only she was the gorgeous brunette in the FHM magazine, I was reading earlier, instead of this drunken Kamasutra practitioner. I struggle free, only to see a puke-covered foot coming straight for my face. Scary stuff. Just gross! In the blink of an eye, I throw my hand up and grab the vile extremity. Meanwhile, my partner, who was indifferent to my struggle, opens the back door. In one motion, I have both legs down and restrained. Thank God for my quick reflexes I learned as a hockey goalie. Now I can relax a little.

How does one have a conversation with someone like this? It's easy, have fun with it. For your own mental sanity, go with it. I go into my alter ego, a Middle Eastern Hindu voice, and respond to her verbal abuse with humor. In the face of her tirade of abusive language, I tell her in no uncertain terms that such abuse won't be tolerated. My partner is now bracing himself on the steering wheel cracking up with laughter, as I feel our office swerve on the road with him trying to keep it on the road. It made us laugh, and therefore made our evening more tolerable.

As we arrive at the emergency room, she tells me how nice I am and that I'm the greatest guy in the world. I tell her thanks and it's my job. We drop her into a bed and head for headquarters. As I think about it, this was Mrs. Southern Comfort's way of dealing with the death of an abusive father. It was her emergency, a cry for help. She tried to drown her depression with a drug that causes depression – alcohol. Ironic and just gross.

Naked LSD

If you want to lose your mind, overdose on LSD. If you want to be certified as a lunatic, run around in a parking lot naked shouting obscenities at every passerby you meet. Tell your girlfriend how much you enjoy oral sex over and over again, and say it loud enough for everyone to hear. If you do this long enough in a public place, they will bring in the modern day psych therapists with their tazers and handcuffs (we know them as the police) and give you the therapy you are so lacking. This is our patient on LSD, and he is defiantly certifiable.

Before I get into the meat of this story, let me tell you about its cultural setting.

Recently, a "crazy man" was tackled by police while acting up. The paramedics checked him out and he seemed fine. He was taken to the police station and later rushed to the hospital and declared dead.

You can imagine the public outcry. The police killed another defenseless crazy person. The ones we let out on our street because we didn't want to provide medical care because it costs too much. He wasn't hurting anybody and the police used force to arrest him for being crazy. Of course, we hear and see all the media hype and don't get to the truth. The truth wouldn't sell newspapers and wouldn't stir up a frenzy of public outcry and sell more newspapers. It wouldn't make people look bad, so let's just leave out the truth. Give the public just enough information to make everyone feel sorry for a fabricated victim. Another helpless victim. Vilifying the evil fascist police and paramedics (I write this with 'tongue in cheek' sarcasm, trying not to forget we are here for the public good). This is the atmosphere we enter as we attempt to rescue the Naked LSD crazy man.

We pull up to an empty parking lot that is circled by about five police cars. A man is face down in handcuffs and his lily-white ass is announcing to the world he has on his birthday suit. He is partially clothed in bark dust, the result of rolling around on the ground after being hit several times by a tazer. He is hollering something about "getting a blow job." His pupils are dilated to the point all you see is the black hole in the back of his mind. The officers state that they were not members of the Greek Olympic team and did not feel comfortable wrestling him to the ground so they hit him with a tazer.

The idea of wrestling a naked man just didn't appeal to them. They hit him with the tazer several times as he was running in circles around the parking lot(drugs often reduce the effectiveness of the non lethal methods police use).

His girlfriend is standing off to one side crying, and saying he took too many hits of acid and she told him not to. The police officer looks at me and tells me they put the handcuffs on while he was stunned from the effects of the tazer and that no other force has been used to restrain him. Remembering the political atmosphere, I look at the officer and tell him we will take Mr. Lilly to the hospital to be evaluated. He is more than happy to help me. We load Mr. Lilly into our ambulance. We are entertained by a string of profanities and a description of the joys of sex en route. I wonder how bad his flashbacks will be. He is only seventeen, and already he has condemned himself to flashbacks of horror that will haunt him probably in the years to come. Mr. Lilly is sedated and left to sleep it off in a hospital bed. This story didn't make the news. It wasn't news worthy. No one died. No crazy defenseless person was hurt. Naked LSD will have flashbacks, but not of news headlines, just naked obscenities, tazers and ambulance rides.

Chapter VI

Babies are Born

There will always be and have always been babies being born. For centuries, women have been giving birth. They have dropped babies in the field and gone back to work without batting an eye. All this has changed in our modern age of medicine. Sure, the survival rate has gone up with birthing rates increasing, and the number 911 has become the favorite for women in labor.

Modern medicine has put paramedics on the front line of baby deliveries. I now have several namesakes running around on the streets because of my efforts. The ladies were so grateful they named their bouncing bundles of joy after me. The thought is kind of ironic to me since most have a Scottish first name and a Spanish last name. Craig Gonzales-Ramirez just doesn't quite sound right to me, but in our society of integration, I guess it was inevitable.

Our stories will reveal the true frontline of babies being born. So get ready to catch.

A Full House

For many paramedics, the scariest call of all is the 'eminent birth' call. We deal with the processes of disease and death on a regular basis. We grow accustomed to this type of call to a certain degree. Sometimes, we get thrown a curve and we have our mettle tested, but generally, we have a routine. Today is a day of testing, testing to the tenth degree. Copy a "code 3, imminent birth." On our arrival, we find a house filled with kids from the age of about twelve and down. I was wondering if the woman ran a day care. It seemed so many kids were in this single story home. There were kids in every room and they were running from each room adding to the chaos. Mrs. Bungee is in the back room being attended by firefighters. As we walk into the room, we see a baby sliding out of the woman like a child in a water slide. The firefighters grab up the kid and she lets out a healthy cry. The cord is clamped and the baby is handed to Mom. She tells us that she is very experienced at this and this is a normal birth for her. I ask the loaded question.

"How many children of these are yours?"

She smiles and says, "All of them."

"How many is that?" I ask incredulously. She looks at the bundle of joy in her arms and tells us, this is number 10. She is, in medical terms, gravida 10, para 10, a perfect 10. No, this woman is almost 6 feet tall and about 280 pounds; maybe not so much a perfect 10, but someone thought she was, the pregnancy a result.

Now for the rest of the story.

When the baby was born, the work is not quite done. The placenta must also be delivered. Usually, this takes about twenty to thirty minutes, affected by the mother nursing the baby. Mrs. Bungee tells me she wants to go to the hospital and to make sure everything is fine with her and her baby. Mrs. Bungee is bleeding some, but not more than usual.

She takes a towel and places it between her legs and tells us she will walk to our gurney. This is a good idea since we can only get the gurney into the front room on account of all the furniture and kids. A firefighter carries the newborn and we assist Mrs. Bungee as she holds the towel. The umbilical cord is hanging between her legs, kind of like a bungee cord hanging off a bridge.

Suddenly, Mrs. Bungee looks at me and says, "Uh oh!"

"What?" I ask.

"I think I dropped something."

At this point, all the schooling I remember about pregnancy comes flooding back to my memory. I forgot to ask if any of those kids were twins. Well, I think if this is another baby, it has come into the world with the experience of a bungee jumper. I look down and there swinging between Mrs. Bungee's legs is a bloody sack. Thank God it is only her placenta and it will be put in a plastic bag and taken to the hospital to be examined by the doctor as well. I check my pants and find they are dry. Another successful call in spite of the *faux pas*.

An interesting note: several years after this event I received another 911 call from this same home for a choking child. When we arrived, the child was fine and as I was holding the child, I recognized the mother as the same lady we had assisted in her child birth. I was holding the two year old I had helped deliver and it was a strange feeling of pride and connection. It gave me a glimpse into the reason I am a paramedic.

Don't Take Care of Me (microwave baby)

If you want to skip this story, you can.

If you do not have a strong stomach and are sensitive to the horrific behavior of individuals toward the innocent, then go on to the next story. I recommend it.

I wish I could get this story out of my head. Maybe if I write it down, I'll be rid of it?

It was a quiet afternoon, a regular day with nothing eventful happening. We call this the calm before the storm. We are often ill prepared for the storm even though we know its coming. We had just taken a little old lady to the hospital for a fever and we were about to get some lunch when the call came down. "Pediatric code 99": this is an adrenaline rush for most medics. It means something is very seriously wrong with a child and they need a "savior."

Today, that is us. We pull up to a mobile home that is at the end of a gravel driveway. It is just outside of the city limits and the woods give the setting a very rural feeling. A dog is tied up in a muddy pen on the side of the trailer and is barking incessantly as we pull up. A truck and a car sit in the driveway with trash piles of different sizes lining the porch.

We grab our equipment and rush into the mobile home.

I am greeted by a middle-aged lady who hands me a bassinet covered by a blanket. She walks into the kitchen and begins to eat a hamburger purchased from the nearest fast food restaurant. She says with her mouth full, "I think it's sick, no one has taken care of it or seen it since yesterday morning." My red flags are waving and the warning signs are flashing in my mind as I look around. The house is a pigsty and a middle-aged man is seated on a sofa with a glazed look on his face. Down the hall, I hear a woman weeping and see the shadow of two people sitting on the edge of a bed. Today, I have a student and I see an opportunity to do some teaching, but I feel a sense of something not being quite right. A sense of being conflicted rises in my gut. We should not be here, is shouting in my subconscious, but I ignore that intuition I have since learned to trust.

I see a bedroom door open in front of me, and a bed, partially made. My partner, student and I move into the bedroom to continue our examination of the baby.

If you want to stop reading now, it's okay.

I will understand. It's okay if you want to skip this part, but I'm sorry, I must continue.

I lift the blanket and under it I see a baby. Well, it used to be a baby. The eyes are completely white, like white marbles. Its skin is a pasty gray with an odd discoloration on the legs and arms. The baby is stiff and dry – very dry. All of its mucosa is gone.

This is the first horror.

The infant's arms are extended out and stiff like it is reaching for its mother, or for me, or for someone to care for it. Its mouth is fixed open and its tongue is dried out like a raisin. The infant is wrapped in another blanket and I lift the child up. The baby is hot to the touch.

This is not right. The baby is stiff, but hot (red flags are flying all over). I look around for a heat source like a heating blanket or warmer but find nothing. The evidence is not adding up.

My partner is going into paramedic mode, a defense mechanism we all develop quickly, and wants to do something. We take out the heart monitor, but I know this is really only a fruitless attempt at a rescue. The monitor shows flatline. Not surprising.

What comes next shakes us all to the core.

A voice from the other room hollers, "Where's the rifle?"

This is the second horror.

We look at each other and know we are screwed.

The room has a small window and the only escape is the door we came in, and that's the same direction as the voice that we would like to avoid. Avoid is too mild a word. The flight mechanism is kicking in and we are trapped like ducks in a barrel waiting to be shot.

My partner grabs the portable radio and asks if police and firefighters are going to be arriving anytime soon.

Then she hollers, "We don't need a rifle in here, we are okay." It sounded nonsensical to us, but it seemed to work.

The lady who handed me the bassinet puts her head into the room and says her husband is still stoned. This, of course, is not comforting until she tells us that he is looking for the rifle to kill that barking dog. It is giving him a headache and he wants to shut it up. Silence the dog with a rifle blast. I don't see the rationale.

We inform her that the baby is dead and that the police are here and a rifle would not be a good idea.

She looks panicked and tells her husband to not get the rifle. She scrambles to put away a bag of marijuana and clean up the drug paraphernalia. The room we are in has several pipes and bongs scattered around. There is evidence of multiple drugs being used in this mobile home. I call for the MX (Medical Examiner) on the portable radio and he states he will be arriving in just a few minutes. He arrives and examines the baby.

"Why is this child hot?" he asks.

I told him this is how we found her and I was wondering the same thing.

I could find no heating blankets or other heat source. The police and the MX interview the uncle, aunt and mother and discover they have been having a drug party.

They don't have any knowledge about how the child got so

cold. After a lengthy interview by the MX, the couple revealed the rest of the story. I was not privy to the conversation so the MX comes over to me and tells me he knows what happened. He asks to speak with me outside. The third horror. He tells me that the baby starved to death because the mother stopped giving the child the right milk. He tells me the mother was tired of the baby crying so she stopped breast-feeding and started formula.

She stated, the baby continued to cry so she gave the baby whole milk. The baby didn't change its behavior so the mother started feeding the baby nonfat milk. This worked because the baby went into starvation mode. Within the last twenty-four hours the baby died, this is when the mother and relations were so high on drugs they did this inconceivable act.

They found the baby cold (and hopefully dead) so they put the baby in a microwave to warm it up, the final horror. I had learned enough. I want to go home now. No matter how high you are, how do you put a baby in a microwave?

I can't remember much after this. I remember getting back in my ambulance and I remember going back to work, but most of the day is a blank. This is a mechanism I am told protects me from psychological damage, but I think it didn't kick in soon enough. I still remember this horrible scene like it was yesterday. Senile dementia might be a blessing one day, but until then, I will remember.

She Hit Me in the Gut, Want to See?

Police are on the scene of an assault. We know this old trailer court. We get called often to this residential center of metal boxes. The labyrinth into the park has become so familiar we easily find the police cars parked in front of the home. We grab our kits and are greeted at the door by a police officer. He points at a lady sitting on a couch holding her stomach. Her 'friend', who is in custody in the other room, punched her in the stomach. I have a student with me today and I look at him and say, "She's all yours, what are you going to do?" He walks up to the lady and begins his interview.

She tells him she wants to go to the hospital and get checked out. She has the strong odor of beer on her breath. She has on a loose t-shirt and sweat pants, both a dingy yellow. Her body odor is the kind that gets in your skin and on everything in your ambulance. This is a perfect setting to see how my student can maintain his composure and take care of his patient, even in the face of such daunting challenges. We load our patient into the ambulance and my student introduces himself and me. Good start, I think. I try not to interfere when my student is learning the art of paramedicine. He does a great job of asking all the questions regarding trauma to the abdomen. His treatment is exemplary. My student has been in EMS for years and decided to get his paramedic. Today, he is going to see the underbelly of EMS. Today, he will be exposed to the… well you'll see. I ask my student if there are any other questions you might ask. He suddenly remembers you must remember the female side of the equation.

He asks his patient, who is almost forty, if there is any chance she might be pregnant. She, in her slurred drunk speech, states she is three months pregnant. We are both caught off guard by her answer. Well, I tell him, you need to ask her the battery of questions this answer raises. He does a good job at asking the right questions, but he misses the question about the possibility of bleeding. I tell him he needs to check to see if the lady might be bleeding. My intention was that she was to check herself. Mrs. Assault was listening in to the conversation and suddenly became very accommodating to my student. She heard me tell him he had to check so she lifted up her hips and dropped her sweat pants, she spread her legs, and for a moment, I thought my student was going to fall into the great void of this woman's vagina. He made a face I wish I could describe in this story, but only a photo could truly capture his expression. It was a mixture of fear, disgust, curiosity and duty. Later, he told me he would like to get his mind washed of the memory of this woman's genital region. To quote him, he told me, "It was the ugliest looking beaver I have ever seen." Needless to say, I was trying hard not to laugh at his introduction to the underside of paramedicine. I told him I was looking for a way to get him, and it presented itself perfectly. I looked at him and said, "Gotcha!" He just smiled and rolled his eyes – he was *got*!

Butt baby

Some stories are just too good to keep to yourself. This one is a classic and needs to be put down for posterity. The eminent birth emergency is very unnerving for most paramedics, and often the simple birth sets the medic on edge. Now let's throw in the not-so-regular birth.

A paramedic is called to childbirth in progress. On arrival, he hears screams in the back bedroom and he takes his equipment to the side of the woman in labor. She is already pushing and panting. She tells him this is her first baby and that she has not seen a doctor, all bad signs. He tells her he must get her to the hospital, but before they go, he must check for crowning. This is the point of no return when delivering a child and usually it means set up camp and deliver the baby. The paramedic takes the pants off of the lady and sets up his OB kit (an obstetrics kit) getting ready for the worst. He takes a look. He picks up the radio and calls the hospital. This does not look right. He has seen childbirth before, but this is somehow very wrong. He gets a doctor on the radio and the entire ambulance company is listening in, you see we are nosy too. He tells the doctor in a very panicked voice that he sees the head, but it's split down the middle. He thinks the baby is deformed or something. She pushes again and out squirts this green fluid. He hollers over the radio that the head just exploded. Oh, my God he says, the baby doesn't have a face. The medic is completely beside himself on the radio, he is yelling something about legs. Suddenly, all is quiet and over the radio you hear a baby crying. The medic, in a very subdued voice, tells the doctor that the little girl is doing fine and that he would like to go home now. The medic delivered his first breech birth. The baby came out butt first and the medic has learned a valuable lesson. Better to be thought a fool than to use your radio and announce to everyone, removing all doubt.

Baby Doe

Tonight is getting hectic and calls are dropping like bombs. My partner and I are on our way to check a patient for the police. Here we go again. We continue 'code 1' through traffic since this is not an emergent emergency. "Breathing problem" comes over the radio and we are closest to the call. The police will get another ambulance since we will be diverted to the "breathing problem." I hit the lights and siren, code 3, since this is a real emergency. Once the call drops, our heart rates increase in anticipation of what is ahead. Half way to the call, dispatch informs us this is not a breathing problem, but a pregnancy problem. My heart rate increases even more. I wonder what is waiting for me at the expecting mother's home? Did her water just break? Maybe contractions just started. Is she just having abdominal pain? Is she going to be delivering a baby? Am I going to be delivering a baby? All these things rush through my head as we approach the house. Upon arrival, dispatch informs us that the baby is crowning. In other words, the baby's head is being delivered as we arrive. I rush in with a fire medic.

When I enter the home, I go down a narrow hallway and turn

into the bathroom. There was Mom, hovering over the toilet in distress. I look down and see half of the baby's head sticking out between Mom's legs. "We need you to move to the floor!" Toilet water is not a good medium for a newborn. Mom says, "I can't!" as I lend her my hand. She reaches for me and her hand is covered with blood. I try to grab her hand but she extends forward and grabs the upper arm of my jacket. Well, I will have to red-bag my jacket because now it is covered in blood. We get Mom to the floor and I get on my knees in preparation for the delivery. It is ironic how almost every pre-hospital birth happens in a cramped bathroom. All this bathroom could hold was Mom, the fire medic and myself. My partner and two other fire fighters are in the hallway, ready to lend a hand. I place my hands on the baby's head and tell the mom to push. The rest of the head is out and the fire medic hands me the bulb suction. I suction out the baby's mouth and nose. At the same time, I tell Mom to stop pushing so I can continue suctioning before delivering the rest of the baby. She relaxes and I tell her we are going to deliver the baby on the next contraction. I finish suctioning and then check for the cord around the baby's neck. No cord could be found, thank God. These are the moments us paramedics get religious. Mom informs me she is ready and she begins to push. I embrace the baby's head and lightly pull downward on the baby's head to help the shoulders past Mom's pelvic bone, and out shoots the little guy. It's a boy! It is very exciting to catch the kiddo after he shot out. I set him down and continue suctioning out some meconium from inside the baby's mouth as the fire medic dry him off. After a couple of attempts at not producing any more meconium, I then rub baby's chest and flick his feet to stimulate a response. He is cyanotic and is not responding the way I would like.

At one minute, his APGAR score was low at four. Mom asked if he was okay and reached down and touched his shoulder. Out came his first big cry and Mom shed a tear. It was like she shocked him to life with a mother's touch. He then became more active and opened his little eyes. He was still a quiet little guy, but he was doing fine. His heart rate and respiratory rate were within normal limits of a newborn. At five minutes, his APGAR increased to nine, a good score. His extremities are still a little blue, but getting pinker with each passing moment. I get up from the floor and have another fire medic step in. I have to get cleaned up. I didn't have time to gown up. I wipe myself off with cleansing wipes while fire medics cut the cord and place the little guy on Mom's chest. He was still very quiet but only because he immediately started to nurse.

We got Mom on the stretcher with baby in her arms and transported her to the hospital of her choice. We bypassed the ER and headed directly up to Labor & Delivery. While taking the elevator ride, I ask Mom if she had a name picked out. She states, "I have about 250 in a book."

I sarcastically respond, "Looks like you got them narrowed down," As we get in a short laugh. The elevator opens and we arrive at L&D. The nurses take baby to the assessment table and we transfer Mom to the bed. I give the nurses a report on the little guy while another nurse asks mom questions. After a couple of minutes, the nurses are doing their thing and I walk over to Mom. I congratulate her once again on her newborn and she squeezes my hand one last time. I tell her it was a pleasure delivering her baby and it was my first.

"Oh, you were the one who delivered him," she says.

"Yes, it was me," I respond. I didn't really care that she didn't know. She was only giving birth. I asked Mom if it was alright to have a nurse take a picture of me holding her son. "Yes, it would be my pleasure," she responds with a smile on her face. I get back to the little guy and ask the nurse to take the picture with my cell phone. I tell her it was my first delivery and she says I have to get a picture with my first. I pick the little guy up and the flash goes off. A picture I will keep forever. It may only be cell phone quality but it is better than nothing. I say goodbye to the unnamed baby and then to mom. A quick wave and out the door I went.

I noticed the adrenaline rush through me, as I had never done something like this in my life. Bringing a life into the world is one of the greatest feelings ever. We see a lot of death in this job and to see the opposite end of the spectrum is a blessing. I had seen a few deliveries in school, but it's different in the pre-hospital setting. I wasn't nervous but rather focused on the task at hand. Women have been delivering babies since the beginning of time. We just went through the motions and delivered a baby. Mom did all the work. A

Chapter VII
Ode to the Emesis

We have to take a class every year that teaches us how to avoid substances that we may be exposed to, go figure, diseases and infections. In this class, we are given packaged propaganda about blood and other body excrement. We are trained how to reduce risk of exposure in these situations. We are to prophetically predict by our astute innate ability an exposure threat. We are to have inherent 'Sherlock Holmes' abilities. This obviously doesn't happen. We still get exposed to all kind of things.

One of the disgusting substances that we find ourselves a target of is, in medical terms, *emesis*, known to most as vomit. The contents of an individual's stomach that is partially digested with stomach acid, saliva and bile and sometimes blood. This is what this chapter is going to be all about, so get your bucket or your paper bag because there is only one word for this topic: GROSS!

We are nauseated by the sight of trivial personalities decomposing in the eternity of print.

Virginia Woolf (1882-1941)

I thought this quote appropriate here. These stories may make you nauseated. So get out your Zofran or Phenergan, or your barf bag.

A Poem to the Nauseated

Taste and see, it is not good
Don't swallow! You already did, yuk!
Bile burns all the way down and all the way up
Around we go, vertigo, I think I'm going to hurl
Here it comes, there she blows, chunks
I drank too much I should have known
It comes out the nose and hit's the toes
I just had this cleaned and now it's ruined
Here it comes my lunch and dinner
Tasted better the first time
This poem really doesn't rhyme
Makes you want to vomit
Who really likes to hurl? Upchuck, spew,
toss your cookies
Dry heave, vomit, emesis, puke, spit up
Not even the bulimic finds it entertaining
Get the mop someone has got to clean it up
And it sure as hell will not be me
By Mr. Hurl

In Your Face!

At the beginning of every paramedic's career, you have a period of waiting for your certification to get in the mail before you can actually practice paramedicine. Before you get your certifications, you have to practice as an EMT-Basic or Intermediate. I was still a basic with the mind of a medic; my schooling was done, I was just waiting for the paperwork to clear.

I'm sitting here with a fully immobilized woman (patient is on a backboard and has a cervical collar on to prevent movement). She was thrown from a horse. She has back pain, but has had it since the last time she was thrown from a horse years ago. Upon examination, she has no other complaints. She tells me that she gets carsick quite easily. I tell her to let me know if she starts to feel sick. Having a patient strapped to a long backboard and unable to move, you have problems if they vomit. Aspiration of emesis is not a good thing. It can lead to pneumonia and death. We are twenty minutes from the hospital and the roads in our county are not straight or smooth.

Midway through the transport, she tells me she is feeling a little nauseous, but it isn't bad. I tell my paramedic partner that she is feeling nauseous and that she could use some medication for nausea. I was implying that I wanted him to trade out and give her this drug. He just told me, "Five minutes," and ignored me. I wasn't very happy at this point. I continued to keep an eye on her. She repeats that she feels nauseous and I get more concerned.

Next thing I know, she says, "I think I am going to be sick." I reach across the stretcher to hit the suction button by the airway seat and it hits me. Thick, sticky, brown projectile vomiting. It splatters the wall behind me and the ceiling above me. My glasses protect my eyes; my mouth is closed, but not airtight. I still taste it. I can feel the pasty liquid around the collar of my shirt, working its way to my under shirt. I gave the suction to my patient and she suctioned herself. I reached for a blanket and wiped my face off the best I could. I then returned to helping my patient. Her first words to me, "I'm so sorry."

Yes, so was I, but I told her, "It's okay, part of the job." I was more pissed off with my partner because I knew this was going to happen and there was not a damn thing I could do about it. I guess I could have gotten out of the way, but it happened so fast. We arrived at the ER and I received some looks from the nurses and doctors. We cleared the call and I returned to headquarters for clean up.

A shower never felt so good. My partner cleaned out our ambulance and I took a nice, long shower and put on my backup uniform. We headed back out on the road because your day doesn't just end just because you get puked on. We grabbed some lunch. This is not unusual for us and we still get hungry. We get another call. We took this patient to the same ER and I get to check up on my last patient. She was given pain medications for a severely fractured pelvis. The ER doctor came by and showed me the Xrays. I will just say it was immediately obvious it was a fracture.

He told me, "You smell better now." And he smiled.

Razor blade puke

Today, we are called to a man who is vomiting blood. Each time he vomits, it hurts. The pain isn't like that bile burning pain, it is different he tells us. Each time he rides the porcelain bus, the rust stains get redder. Throwing up blood is not a good thing.

One thing I have learned the hard way is that patients lie to you. Here we are, trying to save their lives and they give us false information. What is frightening is that we make medical decisions based on this false information. Do they realize this? No! I wonder if the truth is so difficult for them that their house of cards will come crashing down if they tell one paramedic the truth. Is the truth so threatening? I mean, I have people tell me this complete set of symptoms and when the symptoms started, then turn around and tell the nurse a completely different story. I can't figure it out. There seems to be no rhyme or reason for their misleading facts. I even try to build a relationship with my patients, win them over to me, and they still find it in their heart of hearts to fabricate their story. I have started learning from 'Sherlock Holmes' and 'CSI', and have come to the conclusion – the evidence doesn't lie. Look for the evidence and the truth will come out. I have exposed my fair share of falsehoods with this method.

Mr. Razor was one of those patients who almost got away with his lies. Mr. Razor called 911 for 'blood in his vomit'. We arrive and find a forty-ish man vomiting and spitting up blood. He tells us he was eating dinner when he got sick. He has a zip tie baggy he is carrying with him he wants to show the doctor. This seems odd to us, but we go with it. We ask him all the usual questions about vomiting blood. How long? How much? When did it start? And is it painful? He states he has only been vomiting blood for an hour and it has been bleeding since then. He seems to be avoiding our questions. The trip to the ER is short and as we unload him, he breaks down.

"I was trying to kill myself and I swallowed a package of razor blades. I have been puking up razor blades ever since." The idea of swallowing razor blades is hard for us to fathom until we notice in the bag he was carrying there are several floating razor blades.

I look more closely at the baggy and it contains three to four flat razors. We are confounded and take our patient to the ER. We come to find out this is not Mr. Razor's first visit. He is a habitual razor blade swallower. I think he should join the circus and become a sword swallower because that would actually be a lot safer. He went to surgery for sutures and a tube that was placed in his belly, through which he will get his food until the wounds heal.

Good Eats?

Here we are again, sitting in our ambulance on a nice Friday afternoon. It is in the heart of rush hour and traffic is getting bad.

My partner just came back with one of those microwaveable double cheeseburgers from the nearby gas station. I can't believe what this guy eats. Some of the food makes me want to throw up. As he eats his double patty delight, I hear a seizure call over the radio. It's just across the street at a popular restaurant. We arrive on scene before our dispatch can even give us the call.

We enter the restaurant and are led in by the hostess. The restaurant is packed as we pass by people waiting for a table. Every table is packed. People give us the usual stare down as we pass by with all our equipment. We are led to a heavyset man in his late forties, who was eating dinner with his wife when he just passed out and went into convulsions for about thirty seconds on the bench seat.

The table is pulled away and I have Mr. Robin transfer to the floor so we can better assess him. Mr. Robin doesn't have a history of seizures, so judging by his size, I start to think he might have had a cardiac event. He is a little pale, diaphoretic and his face is flushed. I start to ask him questions as we hook him up to the monitor and obtain his vitals. He tells me he is having a little pressure in his chest and that he just remembers feeling dizzy and then blacking out. His EKG showed ST elevation and occasional PVC's. He is having a heart attack.

I ask Mr. Robin if he feels nauseous and he tells me he is going to be sick. My partner states "Turn him towards you." I think to myself, I don't think so. I am between Mr. Robin and the bench seat. There is no place for me to go. I help turn Mr. Robin onto his right side, facing the large area that once contained a few tables. As soon as we turned him, I look up because I know what is going to happen. I see all these people eating away at their meals, some watching us tend to our patient. Their interest is high as it's just like TV! In the blink of an eye, hundreds of people are simultaneously grossed out and immediately stop eating. Mr. Robin throws up a thick pile of emesis over the floor about an inch thick. The contents of the food are the same as surrounding patrons are eating. It is obvious Mr. Robin has eaten a Cheeseburger, fries and a chocolate milkshake. Most people ask for their tabs and scurry out of the restaurant. We get Mr. Robin on the stretcher and then head for the exit. Now more people are at the entrance and they clear the way just like a red carpet. We pass the people that lost their appetites, courtesy of me and my perfect patient positioning.

Upon getting Mr. Robin outside, he tells me his chest pressure is getting really tight and is starting to have a little chest pain. I look at the monitor and now he is in Atrial Fibrillation. His heart is very irritated and we head for a nearby ER, less than five minutes away. I start an IV and give him some Aspirin and Nitroglycerin. We arrive at the ER as he is getting the full workup, and possibly more. I tell Mrs. Robin that her husband is probably having the early stages of a heart attack. I get my paperwork and head out to the ambulance for our next round of fun. I get in our metal box and my partner looks at me, not saying a word. "What?" I ask, "Did you see

Chapter *VIII*

Intermission

When we start this job called EMS, we often find it consuming every moment of our day. This is why I have put this chapter here. We need to take a break. In fact, put this book down now and go do something for yourself or for someone you love.

If you're involved in EMS, go do something that has absolutely nothing to do with EMS. Better yet, take your wife and or family out to dinner. Do this on a regular basis. Make it a habit for your sanity's sake.

No matter how much you love EMS, you must take a break or you will come to hate the thing you love. This is not good.

You are too important to the ones you love.

You are too important to EMS and if you don't take care of yourself, no one else will.

I hope you took my advice and you're now picking up this book to finish.

Let us tell you a story about respect.

A Little Respect

Sometimes, we do get a little respect.

One day at post (we don't have quarters any more, too expensive), I decide to go into Dairy Queen to get a Diet Coke. I end up in this long line, but figure I'd wait. I got time unless a call drops (someone calls 911). Various sized people are in line: there is an obese man that needs to follow "Jared's advice"; a skinny man that stays thin no matter what he eats until one day he dies of a massive MI (myocardial infarction); a short Latino; a couple of tall African Americans; and two high school chicks ahead of me in line. Out of the blue, the cashier, who happens to be the manager says, "You." With his long pointed index finger directed at me, "Come here." I look around and point to myself. "Yes you." He states as he gestures me forward. "What will it be?" he asks as I approach the counter.

Before I respond, a voice behind me asks in a stuck-up tone, "What's up with that? Why does he get to cut?" The manager turns his focused eyes from me to the two high school chicks, with their fancy in-style highlighted hair cuts, painted fingernails, cell phones with awful ring tones and fake tans.

"Do you know how many times these guys order, pay, then

leave without their order to help some prissy, inconsiderate person like yourself?" states the manager with glaring eyes, "We're lucky to have people like this around, and you can leave if you don't like it." The high school chick is silenced. The manager turns back to me and asks, "So, what will it be?" with a smile on his face. I order a small diet coke. He charges me for a small and gives me a large. Ah, the perks of the job. Sometimes we do get recognized for our hard work, but this is an anomaly – thus the honorable mention. Let us give you a little history of 911 before we get on with our stories.

Call 911!

The insanity started with the three little numbers everyone has learned since kindergarten.

That number the firefighter and the police officer asked you to call if you, or someone you know or don't know is hurt and you need help, or they need help.

You know what it is!

It has been pounded into your memory until it has reached your subconscious. Even your pets know the number. Yes, it is 911! I bet your pet could even dial it for you.

I need help; maybe I should call 911?

What was that number? Yes, 911.

Who you gonna call? *Ghostbusters*! (Sorry, just a moment of insanity.)

NO, not ghostbusters – *paramedics*, the saviors of us all! Or, the bringers of death? Angels from heaven or Angels from hell?

Someone to shift responsibility from you and your emergency to us. Again, it depends on your perspective, your reality.

If we fail, angels from hell; if we succeed, the other ones.

The idea is born with a simple dialing of '911'.

That you could pick up a phone and *presto,* we are there to meet your every need. To save you.

Visions of Star Trek come into my head. Beam me up, Scotty! I need to get out of here.

Pick up the phone and in eight minutes, no, five minutes, no, right now! We'll be there. Materialized in the midst of your emergency like Captain Kirk or Spock rescuing you from that ugly monster- your emergency. That's the expectation. No, that is the requirement! The government has made it the right of every citizen and alien alike. The law states it is the duty of every EMT, for that matter every medical person to respond to the call for help. Can you believe it, they made a law that requires us to do what we love to do, maybe they know something about the nature of this job they are not telling us. They have even made a law to protect the person responding. They call it the Good Samaritan law, you know, after the story in the Bible about the Samaritan who helped the man who was mugged.

Well, this has the lawyers running to the bank and the medical personnel running to the insurance companies to take out liability insurance.

Now, we can all help each other and hopefully not get sued.

We will be there – to rescue you from whatever emergency you perceive to be an emergency.

Yes, the definition is that loose; ask any doctor, lawyer, or paramedic, for that matter. They have no clue what an emergency is – isn't that comforting?

What is an emergency?

Who the hell knows? Maybe if we call the 911 operator, they can tell us?

Ring, ring, ring, ring, ring, ring, ring… If this is an emergency, please hang up and call 911.

Dial tone--------------, what the hell, I *did* call 911!

I redial: ring, ring, ring, ring...

Operator answers: *What is your emergency? Is it police, fire or medical?*

I say nothing… silence...

Hello, are you there?

I breathe heavily...

If you called this number as a prank, you will have a police officer knocking your door down, if you don't tell me right now what your emergency is, do you understand?

Hello, hello... *dial tone*------------

I quickly hang up; maybe they didn't trace the call?

Ha, forget them for hanging up on me! Those "sons of mother-less goats" (we are not allowed to use profanity, even if we are so frustrated we want to, so we invent politically correct phrases).

Knock, knock, POLICE did you call 911?

I hear at the door.

BAM.... I fall to the floor as the tazer violently twitches my body.

Man, that was fast, I think as they put the handcuffs on my wrists and read me my rights.

(By the way this really didn't happen to me but it was sure fun to imagine.)

Of course, if you really needed the police or Fire or EMS that fast, they would arrive only after the fact, but that's a different story. Oh, don't get me wrong, it's the system that is at fault here, not the wonderful police officers, firefighters or medics. Our inability to define an emergency has made everything an emergency. The system is so overburdened by all the non-911 calls, it can't keep up, each 911 call must, by law, be investigated, so don't get caught dialing the wrong number. We are all caught in the jaws of that system called 911! Your reality!

Our struggle

We were just sitting minding our own business, drinking our cup of coffee, enjoying the rain beating on the roof of our metal box (our ambulance), feeling at peace within our little world of insanity when we got the call.

Now we get to jump into the mad world of suburban traffic.

When you look in your rear view mirror and see those emergency lights flashing and finally hear the blaring horn and siren, remember my insanity! I have to deal with drivers, add infinitum, like you, who don't see us to get over and let us get on to our emergency, which used to be someone else's emergency.

Someone just called 911 and you won't get "the hell out of my way."

Oops, sorry, I'm not supposed to have road rage, but if I show up two minutes later than I am supposed to, it's my fault. Yes, it's my fault the 350 pound mom is dead, even if she has been killing herself with over eating and smoking for years. It's my fault you slammed on the brakes in front of me and stopped in the middle of the intersection. So what if I eat my steering wheel. So what if my knees are jammed under the dash and my lower back is in spasm.

Well, that's all in a day's work (workers compensation may not cover incidentals, so the knee replacements and the chiropractor are probably going to be out of my pocket). Welcome to the world of EMS.

Yes, I'm supposed to drive around you and smile and *act professional.*

Even as the explicative inundates my mind and my Turrets wants to take over, calling you every name in the book as I inwardly make obscene gestures, knowing I must *act professional.*

So next time you pick up the phone to call 911, remember my insanity and pray it's been a slow day for me because I'm all you've

Chapter VIII
The Crazies

We started this book by quoting Shakespeare's Hamlet. The play is a story about tragedy and a story about insanity. The question throughout the play is what insanity is and what is reality.

The story never really answers the question because the pursuit of sanity is interrupted by tragedy. We as paramedics grapple with this issue almost on a daily basis. We often go from insanity to tragedy and back again. Like Hamlet, we face the question of our existence by being confronted with people in desperate pursuit of the same. The following stories are but a sample of the constant assault upon our sanity by the crazy.

Something happened to me the other day. I read one of these stories to a coworker and all he could say was, "How can people do this to themselves?" All I could say was, "I haven't figured it out either."So if you read these stories and figure it out, let us know.

Enjoy and try to remain sane.

Mrs. Vampire

I seem to be a magnet for all the psychiatric patients. I always get the crazy people. Now that I think of it, just about every person we meet is a psych patient in one form or another. Some are bipolar, others schizophrenic. OCD, PTSD, MPD and depression just to name a few. The list continues with Alzheimer's and paranoid psychosis. There is nothing like answering the same question over and over and over again. Just about every geriatric in a care home has depression. I wonder why, you're 90 years old! I'd be depressed too if I were in a place like that, eating TV dinners made in mass production only to be served cold. These homes have replaced the county mental hospital. These homes are sometimes your neighbors. Comforting isn't it?

We get called to the scene of an unknown psych. Police and fire are already on scene. As I approach and enter the single level home in a residential area, I am stuck in the long narrow hallway leading to the kitchen. The room is filled with too many emergency workers. Sitting on the floor is my patient. She is a skinny 5' 6" tall brunette that has freaked out. Her caregiver says she just freaked out and bit down on her shirt. Then she tugged on the shirt and pulled out one of her bottom teeth. The bleeding was controlled and she has dried blood on her chin, blood staining the front of her dirty white t-shirt. She's really calm, and a firefighter walks her towards me. I introduce myself with a soft voice. She walks with her head down and with a steady gaze. She is blind and mentally challenged,

I am told by the emergency workers. The firefighter gestures this and he reads my lips as I say, "I know." I tell her, "We are just going to head up to the ER to get you checked out." I walk her out the door, down the driveway to our waiting ambulance. She gets in and sits on the stretcher. Away we go.

I tell her I am going to put on the seatbelts for safety. She isn't really saying anything. Then she starts crying again. She asks me, "Where's Samba?" I don't have the slightest clue who Samba is. What comes to mind is the Disney movie with lions, "The Lion King." My first warning sign I completely blow off. I go with it but get nowhere. Then I assume Samba is her caregiver. "Samba will be at the hospital when we get there, okay?" I state with the hope that it relaxes her. She calms down with the occasional tear. I grab my paperwork and her bagged tooth. We head into the ER and inform the receptionist of our arrival of our patient.

"You will be going to triage," she says.

"Triage!" I quickly reply, "She's a psych patient that is a threat to herself; she needs a room."

She replies, "I'll have to get the charge nurse."

We wait for a couple minutes and our patient becomes a little aggravated by the delay. She unclips her waist belt and I put it back, keeping her safely restrained on our gurney. "Where's Samba?" starts to come up again; only this time, it's repeated over and over again. She starts to squirm on the stretcher. We hear 'Room 29' and away we go. Just as we turn the corner, she flips out. Her body lifts off the stretcher, kind of like the exorcist. We stop the stretcher and I go to her side to find the waist belt released again. I reach over and grab hold. I hear it snap and then out of nowhere, I have a bloody mouth on my forearm. It was just like a zombie movie except accelerated tenfold. No slow motion here.

With her clamped down on my arm, I just clench my jaw and hold back any verbal outburst I might regret later. At this point, the entire ER is looking our way. I pull my arm back a little and she releases her death bite. I now have three teeth marks in my arm because of her. My partner throws me a blanket sheet and I use it like a rope, pulling across her torso, around the back of the elevated stretcher. I pull back and it pins her down so we can move her into her room. After we transfer her to the bed, I go straight for the sink to wash up. I have a little blood on my arm from the bite.

I am washing and the nurse looks over at me and states, "Are you going to give report?"

I reply, "I just got bit, wait a minute! She bites by the way, as my psycho turns her head towards the nurse.

"Where's Samba, I want Samba," she continues. I clean my arm and realize I just got exposed to who knows what? Why me? I think. All I was trying to do was help.

After the call, we call our supervisor and I go to another ER to get my blood drawn and tested. The worst part other than getting bit and exposed to who knows what is the fact that I got bit by a retarded blind lady. I am the brunt of every joke at work for at least the next month. Hell, even the lesbians at the coffee shop give me shit about getting bit. Every day, my partner asks someone, "How do you get bit by a retarded blind lady? Ask him," as he points to me. I just take the laughs and go with it. I got bit. It sucked, but it was like something out of a Hollywood horror flick, only real life flesh and blood. Hope I tasted good. I later found out her outbursts have decreased in severity and she got her tooth put back in place. Glad she is doing good, but next time, Samba, tell me she bites! Oh yeah, by the way, Samba is her caregiver!

Mrs. Saddlebags

From time to time, we get to assess a patient at a care facility that is not acting quite right. For whatever reason, the nurses at these facilities think something is wrong with a patient under their care, so they call us. Sometimes it's legit, other times it's just a waste of time and resources. It's a hard call when the patient has mental illness or dementia. Determining if the patient needs medical help is often difficult with these patients. They are usually a case of TMB (too many birthdays), or something like that, but sometimes it is a stroke or medical issue. On this particular occasion, I am greeted by the nurse at the nurse's station. She tells me Mrs. Saddlebags has been hitting her roommate and has tried to bite a caregiver. The nurse gives me the patient's paperwork and tells me she is down the hall and around the corner.

As I turn the corner, I pass a comatose elderly man that spends his day hunched over, asleep in his wheelchair. I then look up and see an elderly woman's face peaking out of a doorway about two rooms down the hall. Her wrinkled face was straight with no emotion, and her eyes were fixed on me. I took another step and out comes Mrs. Saddlebags with her walker and nothing else. Mrs. Saddlebags is naked from head to toe, and her breasts sag down past her waist about halfway to her knees. She hunches over her walker like a sprinter at the starting line, and here she comes. Like a bull fixed on a matador's red cape and I am the red cape. As she "shuffle ran" towards me, her knees punted each breast as she tried to chase me down the hall. She moved quite fast for an old lady and I wanted no part of it. I turned and walked quickly out of harm's way, back to the nurse's station. She followed me all the way to the nurse's station and the nurse came to my rescue. Imagine that, I needed to be rescued from a little old naked lady. I am sure the scene if caught on tape would have been hilarious. The nurse was the only one that could get through to her. She calmed her with words of comfort and direction. Mrs. Saddlebags calmed down and sat on our stretcher as we belted her down. I got a warm blanket and covered her nakedness. The warmth put her to sleep and we got her to the ER for her psychiatric evaluation without incident. I am still haunted by the horrors of the chase, maybe I should get a psychiatric evaluation the next time I'm at the hospital, though I'm sure I would get a room next to hers.

Mrs. Spirit

As we take another patient to the hospital from a nursing home for 'strange behavior', the little lady looks at me and asks, "Who is that following us?"

I look out the back and all I see is an empty street, only the back of street signs flashing by.

"There is no one following behind us at this time," I tell her. Thinking that maybe she thought had a family member following us. Maybe she was expecting someone to meet us at the hospital?

"No look. That person out there is trying to get in." I realize this is her condition that is precipitating the ambulance transport, so I go with it.

I humor her and tell her that the ambulance is protected by a guardian angel and he is keeping us from being tailgated. This seems to satisfy Mrs. Spirit. She sits with a content smile and looks at me. She then says, "Of course, young man. I knew that, I was just concerned that he would not be able to keep up with us. His wings are going so fast. I can hardly see them."

I tell her if we let him in, he won't be able to do his job.

She smiles and says, "Well we can't have that now, can we?"

"No, our guardian angel is a very important member of the team," I tell her.

She looks at me with a confused look.

"I don't know what you're talking about, young man, but I am the one going to the hospital for seeing things. Maybe you should check yourself in too."

I am at a loss of words. Mrs. Spirit called my bluff. Maybe I am as crazy as she is?

Next time I see things that my patients are seeing, I will have to call 911 and get some help.

The Therapist

We are often called to homes that care for those who are 'not all there'. We used to call them insane asylums, now they are PC 'group homes'. Electric shock therapy was a standard practice. We had institutions and hospitals for the insane, the mentally handicapped, the loon, the crazy person, the "One off his rocker," "Out to lunch," "Idiot," "Retarded," "Psychotic," "Sociopath." I'm sure there are a couple more I forget. Now we have PC (politically correct) terms. What in the hell does politics have to do with someone who has lost their mind? People who do not have the mental capacity to care for themselves should not be put on the street because some 'shrink' is more concerned about the 'fiscal responsibility' than he is about the humanity. I think they're crazy to put clinically insane people with sane people. So here we go. I know it's crazy.

We walk into the house for a young woman in the fetal position, curled up in a corner. On the outside, the home looks like your typical residence. The paint is flaking a little. The roof is a year shy of replacement and a couple of windows are duct-taped with cardboard over the panes that have been removed. The porch is creaky as we walk up to the door and knock. A young woman pulls open the door. Behind her is an individual with bugged-out eyes, moaning and drooling down the front of her t-shirt. The person looks like a girl, but her shaved head and facial features makes it a close call. So this is our patient. No, she is in the back room with the therapist.

As we enter, we note the institution has been relocated. There are no pictures on the walls. There is very little furniture and the walls are a dirty white and gray. Each wall has a hole through the sheet rock. Pink insulation protrudes from these holes. The carpet is a seventies shag rug in desperate need of cleaning. The lighting is only a stark glaring light bulb with no fixtures, and a broken pull chain for a switch. I think if I had to live in a place like this, I would go crazy. The smell of disinfectant, urine and bowel movements hang in the air. A young woman walks up to me with only a long t-shirt on and no bra, giving me a lustful look. My partner chuckles and gives me that look that says, "Here's your next date." I give him that, "Don't even think about it," look as we go into the back bedroom. This room is void of all furniture and the walls look like a war zone. There she is, curled up on the floor. Another middle-aged woman walks up to me with an air of authority and states, "This is a girls group home for the emotionally and mentally challenged."

Yesterday, Patty (That is one of the names she goes by) had an episode. She has not eaten or taken care of herself now for more than a week. She will drink but she has not communicated.

The therapist tells us to keep our gurney outside because it disturbs the disturbed. "I will walk her out to you," she says with an arrogant confidence. We stand outside with our gurney, prepared for the insane. The therapist escorts the young woman. She walks out like a normal seventeen-year-old girl and then sits onto our gurney as if she is an experienced ambulance rider. She has done this before. She looks at me and says, "Hi!" Okay, she seems normal enough to me. I will give her the benefit of the doubt. I know doing this sets me up to look like a fool, but what have I got to lose.

"How are you today"? I ask the girl.

"Hungry," she says, "but he food here sucks." Sounds like a normal teenager to me.

"Is that why you haven't been eating?" I ask. She looks at me with that teenage look of "No duh!" The therapist states she will be riding with us and that right now I am talking to Susie. I look at her a little confused. She tells me that Patty and Susie are the names of this young girl's multiple personalities. Later, I find out that there are forty-nine. How do you keep track of forty-nine different people? Let alone their names. I mean how much work is it for the therapist to keep track of all those patients in just one person? As I talk with Patty or Susie or whatever her name is, the therapist keeps telling me the name of the person I am talking to. When we've finally reached the hospital, I have interviewed Susie, Patty, Becky,

Martha, Bertha, Elizabeth, Lizzy, Maggie, and I forget the others. What is amazing to me is that the therapist gave me all these names, but I did not notice one personality change in my patient. Not once did her voice change or her demeanor alter. After we dropped off our patient, I came to realize that the real patient was the therapist. She has all those personalities and was projecting them onto my patient.

No wonder the teenage girl was confused and wanted out of that place. I told my partner we had a multiple patient call and the therapist should have been the patient.

Mrs. Schizo

In our neck of the woods, the EMS system is composed of the publicly funded fire departments and privately owned ambulance companies. The fire departments get our tax dollars and private ambulance companies bill those that we transport. So the next time you have an emergency and don't want to call 911 due to the possibilities of a bill, don't worry. We only bill if we transport, if we respond and check you out, no charge. If you request transport and step into our office, then you get billed. That means if you want to go to the hospital every time you have a cold, you're paying the big bucks to spend time with us. Sometimes, we get what are called frequent flyers. These people call us for every little thing that could possibly be wrong with them – the hypochondriacs that abuse the system. This brings us to Mrs. Schizo.

I find Mrs. Schizo sitting on a dining room chair. The call originally was for a breathing problem, but things change. The firefighters state she has a cold. They are standing back from her, not from her sickness, but because they just want to load and go. This means that we just get the patient on the stretcher and transport. Our assessment is limited to a visual check and a few questions. In this case, there isn't much to assess so we just go. I help Mrs. Schizo to the stretcher and she shakes her head "no" when I ask her if she is having breathing difficulties. She attempts to speak. Her words are stilted and it becomes obvious she is deaf. She is really hard to understand. I then ask if she is hurting anywhere and she points to her stomach. This changes my line of questioning as abdominal pain is more serious than her cold. I attempt to ask the one question every medic should ask a woman of childbearing age, "Are you pregnant?" She nods, yes.

Further questioning results in my own confusion over what is going on due to the communication barrier. I get a note pad and write, "How long? First time being pregnant? Any complications? Prenatal care? How bad is the pain? Sharp or Dull?" She points to the questions and says, "Long." as she shakes her head no. She follows with subsequent head shakes and nods with incomprehensible words. I end up getting the following information: This is Mrs. Schizo's first time being pregnant, she has not seen a doctor, and has a sharp "8/10" pain in her abdomen. She doesn't look to be in a lot of pain though. She sits still, caressing her belly as if it were a live child. I advise her I am going to start an IV and she says, "Okay," elongating the word. After I secure the IV with tape, I hear her cough a couple of times. No big deal, didn't sound too bad. She sounds a little congested. She denies having trouble breathing again. Although it's like digging a needle out of a haystack, I feel good about getting some information with the communication barrier, and the fact that I have decent information to report to the nurse when we arrive at the ER.

When the nurse arrives, I start to give her report when Mr. Schizo comes walking in. I had seen him on the scene, but he didn't say more than, "See ya at emergency," and off we went. I told the nurse; Mrs. Schizo has had a cold the past few days and denies any breathing difficulties. Upon further questioning, she complained of abdominal pain. She also states she is pregnant, no more than a few weeks. As I state this, "She also states she is pregnant," part, Mr. Schizo states, she isn't pregnant. I look at Mr. Schizo and say, "What?"

He tells me, "She isn't pregnant, she always says she's pregnant. She's just sick in the head, schizophrenic." It is perfect timing since the ER Doctor just walked in and heard everything. I felt like a complete idiot. I forgot that patients lie to you and I was duped. Here I thought I was doing so well receiving what information I got, but now it's all for none. I got no information and I was lied to. Maybe I wasn't talking to Mrs. Schizo. Maybe it was one of her many personalities. I wonder how many she has. Who the hell was I talking to, it definitely wasn't Mrs. Schizo! I left the ER Room after completing my poor excuse of a report, and met up with my partner walking back to our waiting office outside. He looks at me a referring to my recent run ins with psych patients and states, "The blind ones bite you and the deaf ones lie to you! What is it with you?"

I look over and state, "Shut UP." with the look of embarrassment on my face. He laughed and I accepted the torment. It was a good laugh and the ER Doctor thought so too.

Bipolar tazer

Most people don't know what kinds of situations we deal with on a daily basis. They see the junk on TV and think they know what it's like. They don't show you the real life situation, just the glorified Hollywood-ized junk that gets ratings. Not every call is the one you see on TV. Most of our calls are really basic, ask questions, start an IV, get to know your patient, monitor them, teach them a thing or two, and learn from them, and then on to the next one. However, some of these calls, although basic, sometimes involve a different twist. Ride around with us for a night or two and you are bound to go on a Psychiatric Emergency, an emergency of the mind. When your mind has an emergency, call the cops and the paramedics. We are called to assist the police. Our job is to help the police and assess the suspect, our patient.

On this day, I am going to assess Mr. Cuffs. He was at home

and got in an argument with his wife. His wife left and called 911 since he was supposedly going crazy. I arrive to find Mr. Cuffs lying prone on the floor with his hands tied behind his back. The cops tell me he has been off his medications for a week and has procrastinated on picking them up from the drug store. He needs them for his Bipolar disorder. When they entered his second floor apartment, he charged at them and they tazed him three times. He had ripped them out and one of the wounds was bleeding just above his right eye brow. No wonder Mr. Cuffs is pissed off. Mr. Cuffs is a different person off his medications. It is like his mind is stuck on crazy. I tell him my name and who I am, although I know it's going in one ear and out the other. He keeps repeating over and over again, "You will listen to every word I have to say. If you don't listen to me, there will be consequences and with this, you have problems." *Okay, this is going to be fun*, I think.

The cops ask me, "How do you want to get him out of the hallway and out of here?"

I tell them we should put him on a backboard and carry him out after tying him to the backboard. They look at me like I was stupid, and I reply, "Do you have a better idea?" They wanted to walk him down, handcuffed. I told them, "Fine, you guys walk him down but first, let me put an oxygen mask over his face so he doesn't spit all over you." They walk him down. It works this time. I think the cops just didn't want to carry his ass down the stairs. The risk was worth it to them, but from my experience, I try to minimize the risk.

After we get Mr. Cuffs on the stretcher, we now have to uncuff the untamed beast. Mr. Cuffs is a big guy and a crazy guy at that. This was the problem that I wanted to avoid. I wanted to secure him one time and one time only. Now we have to uncuff him and then tie all extremities to the stretcher. This is the risk factor when dealing with these kinds of patients. The same goes for getting him off the stretcher in the ER, more risk. We uncuff Mr. Cuffs' arms. I have one arm and a cop, the other. Mr. Cuffs explodes in a rage as I step back clear of harm's way. A second later, this 275 pound man with a deep monotone voice screams like a little girl. The officer at the foot of the stretcher whipped out his tazer and hit him again with the paralyzing electricity, the non-lethal force cops are adopting as an alternative. It's really intense, and almost comical. Anyways, the cop hit him for a few seconds and after that, the cops secured his arms and legs while he recovered from the shock. I just stood back watching. Mr. Cuffs has his eyes glued on me as he hyperventilates like a bull in anticipation of the charge. I hope I don't have red on.

In the back of the ambulance, Mr. Cuffs is keeping fairly calm. His white skin is bright red with silent rage. He wiggles around and tries to contort his extremities for leverage against me. My partner is driving as fast as he legally can, seeking a way to reduce the risk, to get this guy out of this cramped high-risk space. If by chance he breaks free, I have my drugs to chemically restrain him, but that takes time and this is a big man. I take Mr. Cuffs' vitals and just make sure the restraints aren't cutting off circulation to his hands and feet. Out of the blue, I bounce up and down on the bench seat beside the stretcher. At the same time, Mr. Cuffs is elevated off the stretcher only to be held back by his restrained arms and legs. With his arms behind, I see a good 8-10 inches of space between his back and the stretcher.

"What the hell was that?" comes straight from my mouth.

"Sorry, just jumped a curb," my partner replied. I look back to see Mr. Cuffs lying back on the stretcher in a normal position. He is now yelling at me, "I'm going to kill you! I'm going to rip your head off! I'm going to kill your mother!"

"Really!?" I reply, "How are you going to do that? You seem like such a nice guy!" He repeats this all the way to the ER. I am sure if he were to break free, he would try to kill me. That is why the cops have our backs and we have theirs. I hope Mr. Cuffs gets his medications. One of the things you learn in this occupation is how to deal with patients who have lost it, and to not take it personally.

Chapter IX

Contusions and Lacerations and

Trauma, Oh My

We are not in Kansas any more, Toto. It's no longer a yellow brick road. It is now covered in blood and body parts. Smashed cars and skid marks. The Wicked Witch of the West is in control of every slick road and traffic light, and she wants death and gore to reign. The Wizard of Oz is an airbag that will blow up in your face, and if you grab onto the steering wheel real hard just before impact, it will break both your wrists. No, this is not Oz.

Road rash is a condition acquired by motorcyclists as they slide along the yellow brick road, and "airbag" rash is a sign of some safety engineer saving your face – both sting like a mother f**ker! (So I'm told.) Especially as the doctor scrubs the wounds with a brush and soap and water so the wounds won't get infected.

Trauma in EMS is considered a naked sport. What this means is, don't get caught with dirty underwear. If you see a paramedic coming at you with trauma shears (usually a florescent color), set aside all your inhibitions. You are about to be naked in public.

Paramedics don't know what zippers are and they definitely don't try to remove your clothes the normal way. A zipper cut is what we do when we cut down each side of your clothes then lift and pull, voila exposed patient. These next stories will expose and reveal trauma in all its gory glory.

The ladder and the tree

The wind had taken the limbs off the poplar the night before, and a large limb hung dangerously over the shop. Someone had to get up there and cut it down. To hire a professional to cut the limb down would cost too much, so for the sake of a buck, out comes the chainsaw and the ladder. This is the setting for disaster. A ladder precariously placed against the tree, about fifteen feet up, then add the additional ten feet to the fall, for the tree sits on a hillside. Now comes the physics problem. How do you stand on a ladder twenty-five feet above the concrete driveway and balance yourself, while cutting the limb using a twenty pound chainsaw, and not find yourself free-falling twenty-five feet?

I wonder if 911 has any physics background?

Maybe if we can call them and ask, they can solve this problem for us? I doubt it.

No, forget the math; who likes math, anyway. Let's just try this acrobatic act, this amazing feat of balance and skill, a circus performance to top all performances. Let us see if we can entertain the neighborhood. The crowds gather and the lights circle this three-ring circus, but wait, those spotlights are red, blue and white. What is this? Someone has called 911. The amazing balancing act went awry. The ladder shifted and as the chainsaw cut through the limb, the performance changed. It did not fall as planned. It cracked and the large limb swung like the 'mighty Casey at bat' and struck the ladder, sending chainsaw and acrobat into a new performance. The balancing act becomes a contortion act. Mr. Physics for a moment defies gravity and comes face to face with reality.

Gravity hurts. Mr. Physics' leg slips through the ladder and becomes wedged by the force of the tree limb collapsing the ladder like a contorted DNA strand. Now for the fall; As the ladder collapses, Mr. Physics pitches forward headfirst into the great expanse of twenty-five feet of free space.

Suddenly, the ladder pinches his left leg and his freefall slows like a bungee jumper and he is placed on the pavement with the skin of his leg curled around his ankle. His left leg is degloved. Everyone gasps. The performance was amazing, but for all the wrong reasons. No one applauds. They all grab their cell phones, leading to a plethora of 911 calls. The sight of a lower leg stripped of its skin is not a pretty sight. Mr. Physic's lower leg has been degloved. He hears the screams and cries and would like to take a bow, but the pain is disorienting.

In fact, he now realizes the screams are coming from him. We arrive and find an army of firefighters working on Mr. Physics. His leg is being dressed with large dressings and he is being placed on a backboard. We give him pain medication and he realizes that he has been severely hurt and his performance will not have a curtain call, but he will survive. Our trip to the hospital is lengthy due to traffic and distance. We tell our story to the doctor and I watch as the doctor puts his hand into the wound of Mr. Physic's lower leg. I think, *Gross, doesn't that hurt?* Mr. Physics is so medicated, he doesn't feel a thing, Lucky him! I now know it would have been a lot cheaper if Mr. Physics had gotten professional help.

Well, he is getting it now and it is going to cost him. What is sad is he still has to get the tree limb cut down, hopefully he will get professional help.

I cut myself in Half

Why is it, the idea of putting a ladder on a tree and climbing it with chainsaw in hand is so appealing to the public?

The firefighters are sitting in their lawn chairs out in front of their station just enjoying a cup of coffee and the fresh air when a gentleman runs up to them with a bloody towel on his neck yelling, "I cut myself in half!" They quickly call for an ambulance and sit the man on the bumper of the fire engine. He is pale and hyperventilating. He has a panicked look on his face and proceeds to tell the firefighters he was up on a ladder using a chainsaw to cut off some dead limbs. He said the limb started to fall and the chainsaw kicked back on him. He said he felt the blade hit him in the neck and cut him across the chest. He ran into the house and grabbed a towel.

He didn't call 911 because he was so close to the fire station so he ran over to them.

We arrive to find Mr. Jigsaw standing in the fire station.

He walks up to our gurney and says, "Please hurry; I need to go to the hospital, I cut myself in half." I try to calm him and ask him if I can look at the wound. Mr. Jigsaw is six foot five. He is big enough to be two men, so cutting himself in half wouldn't be so bad, I think to myself. As I remove the towel, Mr. Jigsaw proceeds to pass out. He was so sure he cut himself in half, he fainted when he saw the blood on the towel. What I find almost comical is that Mr. Jigsaw had only little cuts on his neck and on his abdomen. Almost like cat scratches, just superficial cuts. Not a single cut is suturable, but Mr. Jigsaw was absolutely sure he had cut himself in half. We laid him down and when he woke up and I told him that he was going to be okay. He did not cut himself in half. I think the experience was humbling for him and maybe he learned he wasn't invincible. We took Mr. Jigsaw to the ER just so he could get himself together. When he arrived, he came to realize he did not cut himself in half.

You could see the expression of embarrassment and relief on his face. He thanked us for our help and was soon discharged from the hospital ER.

Timber and the long walk home

Trees can be dangerous. They can be beautiful and they can be used for all kinds of products that make our life better. Today, we are going for a walk in the woods. Today, we get to see the dangerous kind of trees. They are beautiful woods, but we have a constant drizzle. Today, we are not going to get to enjoy the sights and sounds. Today, we get to discover that the walk in the woods is arduous, and a struggle to save a life.

A lumberjack was deep in the woods cutting select trees for harvest. His job was to walk into a remote location, cut down the tree and mark it for removal. Today, the plan has not worked out. Jack has made a bad calculation on a tree and it did not fall as he had hoped.

A lumberjack hears the cry for help in the woods and fortunately, he has a radio for communications. He radios his boss and lets him know there is an emergency. Lumberjack down! We are twenty miles away from the logging road. The logging road is five miles of winding turning treacherous road that is a mixture of gravel and muddy clay. As we drive this long winding road, we hope the road will take us close enough to rescue the lumberjack. We are flagged into the tall stand of timber and drive up a poorly developed road. A lumberjack meets us and tells us we will have to walk in. We collect the items we might need to help the injured man.

I ask, "How far is this going to be?" The guide tells me it is about thirty minutes in, over an undeveloped trail. "How big is this guy?" I ask.

The guide looks at me and tells me he is no small man and that he has a broken leg. We begin the long walk, and as we walk up several ridges and over several hills, I begin to doubt our ability to affect a rescue. How are we going to carry a big man out of these woods between myself, my partner and one guide? Just as I am about to question the logistics of this rescue, we hear voices coming from the woods. We step up the pace and after several more ridges, we find a man being helped by a coworker. I try to reach dispatch on the radio to no avail. The lumberjack attending to the fallen comrade tells me the radios don't work here.

"But if you go up on that ridge, they get out," he tells me. I look up at the ridge he has pointed at and feel completely exhausted. I just walked thirty minutes up hills and through dense woods and now I have to climb more so my radio would get through. The lumberjack recognizes my frustration and tells me that they know where we are and that the helicopter can't reach us. He then tells me the good news. There is an undeveloped road at the base of the hill and we can use it to get our patient out. The road is flatter, but it will take longer to get back to your ambulance. We put Mr. Casey on the backboard and look at his leg. He has Kevlar chaps on, but they did not protect him from the force of the tree limb. Under the chaps, his femur has exploded and stuck through his thigh. The chap is lined with adipose tissue that burst from the blunt trauma. Mr. Casey is hurting a lot. We give him some pain medications, which take the edge off, but this injury will hurt no matter what, the medicine will only take the edge off. We begin the long walk home. The four of us each take a corner of the board and carry him down to the road.

This was no easy feat, the hillside was muddy and the leaves were like walking on ice. We walked, and we walked, and we walked for over an hour carrying Mr. Casey. We arrived at our ambulance muddy and tired. Mr. Casey was taken to the ER and surgery. He recovered after
weeks of rehab.

The kicker was we had spent almost four hours rescuing Mr. Casey and we still had over six hours left in our shift. At the end of our shift, we both went home exhausted, but satisfied we had effected such a difficult rescue.

Traumatic Dildo

Our society is changing. The things we once knew or took for granted are coming into the light of society's view. Some of this is good, some not so good. Privacy lines are blurring to the point that there is nothing private. So we make laws to protect our privacy, like HIPAA, and muddy the waters even more. But for the media, all is fair game, subject to being brought out into the public light and scrutinized, judged and displayed for all to see. Some of these things we want to stay private.

I see no benefit to society or to our social mental health to

135

display some of the things we have 'let out of the closet'.

Let me explain with the following story.

A call comes down for 'vaginal bleeding'. Now all kinds of things can cause vaginal bleeding. Usually, a bad period, a miscarriage or pregnancy problem can be the cause of this problem. Not today!

Today, we are to be educated in an area we want to keep private. As we walk into this nicely kept apartment, we see a photo of a middle-aged woman and a young woman, on the coffee table. We assume, daughter and mother. Another odd thing happens. The firefighters, who are usually so anxious and willing to help, seem to be exiting stage left. There are four of them and they all are volunteering to go out and get our gurney and any equipment we night need. Their help is appreciated but a red flag goes up in the back of my mind. Something is not right. They tell us our patient is in the back bedroom. My partner today is a female, which makes this a little easier, but not today. We walk into the bedroom and find a woman in her early thirties, lying on her back with a bloody towel between her legs.

My partner asks, "Why are you bleeding this much?"

She looks up and begins to explain that her girlfriend was using a large sex toy to make her orgasm. When the device, which was rubber on the outside and metal on the inside, broke, it cut her vagina.

My partner asks, "Can I see this device?" The woman refuses

to show her or tell her where it was. Privacy still has a place if you're trying to protect your dignity, I guess.

She said she has been bleeding for about twenty minutes now, and is feeling weak. It was at this point, my partner has a moment of realizing the gravity of what was going on with the firefighter's social dilemma. She looked at me questioningly and with a little authority said, "You're not going anywhere?" A little homophobia was in her voice, I thought, maybe it was just me, or maybe she didn't want to be left alone trying to attend to a bleeding lesbian patient. I reassured her I would be there for her, and I told her I would start the IV so we could get her going to the hospital. She appreciated this and off we went. My partner attended the patient as I drove.

We were close to the hospital when my partner said the patient's blood pressure was dropping and asked If we should go to the trauma doctor with this patient. We have certain rules to follow to make the decision to enter a patient into the trauma system. I ask what criteria we would use to make that decision. She tells me, "Penetrating injury." I almost drive off the road. The idea of the patient having a penetrating injury is absurdly funny.

I could see the headlines now. "Lady found with penetrating injury from giant dildo." I'm sure the lesbian community would definitely be "up in arms."

"Well," I said, "Why don't you call the hospital and get some advice?"

Now I'm driving and thinking, "Thank God I don't have to make that radio call!" To call the hospital for a consult with a doctor when everybody and their brothers are listening in was scary. So much for privacy. My partner did a very professional job with a difficult situation. I learned a valuable lesson with this call. If ever I have a patient with a dildo injury, I'll know what to do. Another thing I can thank God for, since that will probably never happen again, but with this occupation, who knows?

The Spice of Life

To be fair, our next story will address the male version of the previous story.

We are called to an apartment with a Middle Eastern gentleman who is complaining of groin pain.

We are dispatched code 3, as usual knowing that this situation is probably not a life threatening emergency, but again the definition is vague. We pull into an apartment complex and drive through the maze, looking for the individually numbered apartments, attempting to match the number to our dispatch's information.

We turn the corner and we see a fire engine parked in front of the apartment. The gentleman lives on the second floor and we leave our gurney at the base of the stairs. We are met at the top of the stairs by our patient, who is holding his groin and squirming like a worm on a hook. He tells us that he was cooking with spices from his country, when he went to relieve himself the spices got on his penis. He is now "in very much pain" and would like to go to the hospital.

He says he has tried to wash off the spices but the water is not working. There is no evidence of his cleansing, so I look at my partner and tell him I will go find out the name of the spice. Something seems amiss, and I go into my detective mode. The apartment is bare except the kitchen, which is cluttered with makings of Middle Eastern foods and curry odors that fill the air.

Well, I think the spice story may be true, but I decide to investigate further. The firefighters are clueless to what is going on and they have found no evidence of anything that might disprove Mr. Spice's story. I go to the scene of the crime, the bedroom and find a mattress on the floor and a pile of adult magazines with several open to attractive nude women displaying their wares. The bathroom is clean and there is no evidence of spices or other accoutrements. I tell my partner what I find, but we see no reason to suspect Mr. Spice's story. Mr. Spice gently sits on our gurney, wincing in pain over each bump in the road. We turn our care of Mr. Spice over to the nurses and tell them the story and they quickly usher us out. My partner goes to do his chart and my curiosity is too much for me. I go to find out more. I see the doctor come out so I approach him to discover the rest of the story.

Now here is the twist (oops! another pun, sorry I couldn't resist). Our patient is going to surgery.

No, it wasn't the spices. Our Mr. Spice was spanking the monkey. Masturbation was his spice of life. He was polishing the rocket so vigorously. he wrapped his testicles around each other. It wasn't cumin on his penis.

Let me tell you about testicular torsion; this is a condition football players sometimes get when they are hit too hard on the football field. These hits are so intense, the football player has his testicles wrap around each other. This is a medical emergency because if the testicles are not untangled, they will cut off the circulation and die; also known as castration. Mr. Spice would not find the spice of life if this condition persisted. Polishing the rocket would become only a memory. Surgery saved his balls, figuratively and literally.

When I discovered this information, I shared it with my partner. He crossed his legs and empathized with Mr. Spice with a grimace and a chuckle. All we could think was Mr. Spice was awfully aggressive in his pursuit of the spice of life, and it earned him a trip to the ER and surgery. Lesson: be gentle in your pursuit of the spice of life, or spanking the monkey might not land you in jail for animal abuse, but it could get you a trip to the operating room.

The flying Mexican

Some people think they are Superman. The man of steel was faster than a speeding bullet and stronger than a locomotive. These people I mentioned aren't. I know it's bad grammar, but bear with me. The way you learn to fly like superman is on a wet, rainy day, when it is dark, dress in dark clothes and walk out into heavy traffic. If you are superman, you will fly, if you are not, you will most likely do what my patient did – crash.

Another cultural phenomenon, I have observed is the complete disregarded by some for cross walks. You see it all the time, kids walking across four lanes of traffic like they own the road. They expect everyone to just stop and let them cross. If you go to Juarez, Mexico, you would understand what I'm talking about. The rules are different there, and when you try to translate that into our culture of rules, you get flying Mexicans. Not to say that Americans don't jaywalk, it just seems that the activity has to increased with the increase in the Mexican population we seem to be presently having. What a great country we live in. Come to America so we can run you over. We don't stop in America. This clash of cultures has some consequences – auto-pedestrian accidents.

An ambulance has witnessed a flying Mexican. We are called in to help. It is wet and rainy, and as we pull up to the scene, we see firefighters and ambulances are wrestling with a young man in his twenties. He is combative because he has a head injury. His arm is fractured above the elbow and his leg is fractured below the knee.

Definitely not invincible! He has a large lump on his head, and he is completely disoriented. The car that hit him is pulled off to the side of the road and the windshield has an imprint where the

Mexican's head made contact. As I viewed the scene, I noticed my patient's shoes were about 50 to 75 feet up the road. This was the point of impact. He did fly! Maybe for a moment, he was superman. I don't think he will remember his moment of invincibility. His head injury will give him the blessing of amnesia. He won't remember the pain either. He might not even remember the ambulance ride.

Maybe someone could at least remind him that if you walk into traffic with dark clothes on, on a rainy day, you will fly like superman but crash like a rag doll, and you will get seriously hurt.

The flying biker lady

Do motorcyclists have ejection seats? Every motorcycle accident I have responded to seems to have motorcyclists flying off their bikes. This usually leads to all kind of different injuries. A particularly memorable motorcycle accident I can recall was a biker couple who were riding together on a Harley. They were in 'stop and go' traffic, and this is very frustrating for a biker. It means you have to keep putting your feet down to balance your bike as you come to a stop. Do this repeatedly and you know you're getting set up for some type of injury.

This frustration increases exponentially if you have a passenger. This is a set up. The traffic begins to move and the biker puts the pedal to the metal. A motorcycle can go from zero to fifty in several seconds. Let me tell you about speed and mass. This is a simple physics problem, so don't panic. Velocity squared, which simply means speed times itself, times half the mass (meaning weight divided by two) equals energy. (V_2 x m/2 = K.) This also equals flying biker lady, because an object in motion continues in motion until acted on by an outside force. Example of an outside force: the bumper of the car. As the motorcycle accelerated and the car stopped, the impact propelled the passenger up and over. We arrive to find a large woman sitting in the middle of the road about 50 feet from the accident. Firefighters are attending her and she is complaining of back and tailbone pain. I walk up to the firefighters and ask them how we can help. They tell me to get my gurney and a pillow. They want to put the biker lady on the pillow on my gurney. I ask them how the lady came to be so far from the accident, in to middle of the road. They look at me like I am speaking a different language. I turn and start walking back to my ambulance. I tell my partner to get the immobilization equipment and I go to talk to the man standing by the motorcycle. Mr. Biker dude tells me he couldn't get the bike stopped and his girlfriend flew over the top of him when they hit the car. Did she land on the car? I ask. No, she flew over the car and landed where she is presently sitting. Unless she had wings, the hard pavement was impacted by the full force generated by her flight. I tell the firefighters this and we immobilize her and take her to the trauma center. She will be needing that pillow, because she had a fractured tailbone and lower back. Her recovery will be slow and she will probably never ride a motorcycle again.

Her lesson in physics was personally a pain in the ass.

Bales of hay and away

Mr. Plant was a farmer. His father was a farmer and his father before him was a farmer. The family farm had modernized with each generation. The farm was producing enough food to feed his livestock and still sell several truckloads of hay to his neighbors. The farm had a good year and the load was so large, they had to bring in a semi truck with a 40 foot trailer. Most of the load was put on the truck with a forklift. The equipment was not quite up to the job so some of the load had to be moved by hand. The bales of hay were between 60 to 100 pounds each, and were held together by bailing twine. Mr. Plant had worked all day loading the truck, and his help had retired for the day while he finished tying down the 15 foot stack of hay. This is the setting for disaster. Mr. Plant is loading the last bale when a piece of twine twists around his glove. As Mr. Plant swings the bale around, the hay he was standing on gave way. The bale pulls him off the trailer and he plants facedown onto the pavement below. Stunned, he gets up and staggers to the farmer's house. His wife finds him semiconscious on the pavement and calls 911.

As we pull up, fire EMS is placing Mr. Plant on a backboard and calling the trauma center. I exam Mr. Plant and find him with coon's eyes (two black eyes). Blood is oozing out of his ears and

144

nose. His upper jaw is instable indicating a "La fortes" (fracture). This is a very serious injury. The bones in his face have broken and his nose is broken. We start two IVs and go code three to the trauma center. Mr. Plant has done a face plant. The bales of hay almost cost him his life. He was hospitalized for several months and the load of hay was delivered to his neighbors. Mr. Plant is very lucky. His injury could have been fatal. He will have sinus troubles for a long time to come. Sometimes, trying to do it alone is not the best of ideas.

Semi with a shoe

This is a story about the incredible. Some would call it a miracle. Some would call it dumb luck. I think maybe it is a little of both. If this young motorcyclist was a betting man, he should have gone out and bought a lottery ticket.

We get a call out into the country for a motor vehicle involving a motorcycle and a log truck. When we hear this call come down, we know that we are probably going on a trauma death. The drive gives us time to get mentally prepared for the worst.

It takes almost half an hour to get to the accident scene. We call ahead and have the helicopter ready to fly prepared for the worst. The drive is up a winding narrow country road with no center lines of paint marking the road. As we pull up to the scene, we see a log truck across the road.

We pull our ambulance to the side of the road and jump out

145

with our equipment. As we walk around the log truck, I notice a tennis shoe hanging from the center axle of the semi. I think this is going to be bad. As we walk around the front of the truck, the impact of the bike is seen on the front bumper. The grill of the truck is crushed in and imprinted with the bike handle bars. The motorcycle is in pieces at the side of the road. Several people are standing around looking at the bike, and attending a young man holding his shoulder and a helmet. I wonder if this is the rider. I look down and sure enough, he does not have shoes on. I wonder where the other shoe is?

The young man is shaken up, but where is all the trauma? He should at least have some broken bones. He should have serious internal injuries and be going into shock by now. The truck driver comes up to me and tells me he was backing into the lumber yard with his logs when this motorcyclist came shooting around the corner. He saw the truck and lay down the bike, sliding under the front of the bumper and then flying up into the grill and bouncing off onto the side of the road. Okay, I think he was probably going about 30 mph when he hit the truck. Ouch, I think, that had to hurt. I go over to the biker and find he has some scrapes on his shoulder and leg, but he is walking around his bike cursing at himself. I ask him his name and he tells me Luke. (Not his real name, but it sounds like luck and I thought it appropriate.) "Luke, how old are you?"

He tells me seventeen, and that he has called his dad who is going to be here soon. I have my partner take Luke's vital signs, which come to find out they are all within the norms. I tell Luke he should go to the hospital and have the road rash cleaned up. He says he will wait till his dad arrives. His dad is a very pleasant fellow and remains calm as he walks up to his son. He surveys the incident and gives his son a hug. He tells him he will take him to the hospital. I tell him that I recommend a ride in the ambulance just to be safe, but he insists he wants to go with his dad. I understand. To have a dad that is more concerned about him than the bike was a refreshing relief. He signed the necessary paperwork and we leave to go to our next call. My partner says he was lucky. I think it was more than that, but say nothing. Pray that you are so lucky if you have a head on collision with a log truck.

Bedpost lady

It's time to put that box of Christmas decorations away. "Traditionally, these boxes of precious memorabilia are put in the attic, but getting the husband to do this is the trick. So if the lazy oaf of a husband won't do it," I will do it myself." Thus the stage is set for disaster. The call comes in as a traumatic injury. This ambiguous announcement is of no help. We are not mentally prepared and are being set up. Little do we know? We pull up to the ranch home and the fire truck is sitting out front with its lights flashing. We bring in our gurney and find a woman lying on the living room floor holding her leg. Our first false clue. The firefighters are attending to a wound on her leg and she is grimacing in pain.

So the question is how did this happen? The husband, who is mumbling about his wife being independent and having to do everything herself, leads me to the garage. The scene of the crime.

The garage has been converted into a bedroom and is cluttered with clothes and bedroom furniture. In the middle of the room is a metal bed. The foot of the bed is made of metal pipe, which is usually decorated by a cap of some sort, but it is now missing. In its place is a piece of human tissue- the plug of flesh from the lady in the living room. The bedpost is covered in blood and a pool of blood is at its base. The firefighters are looking over my shoulder and making faces of disgust. I think, what do I do with this? I ask the firefighters to get me a plastic bag, and with gloved hands I remove the tissue. It is a little difficult to get out, but when its suction releases, I find a piece of bone that is spiral shaped. I think this may be part of my patient's femur. I ask the husband to tell me what happened. He told me his wife was putting away the Christmas decorations when the ladder kicked out from under her and she fell onto the bedpost.

"I heard her screaming, so I came out and lifted her off the post and brought her into the living room." (This is our second false clue). I tell him we need to take her to the hospital and get some pain medications for her leg. We take her to the local hospital and go back out to answer another call. Later, we find out the truth from the nurse. She tells me the patient was transferred to the trauma center for major surgery. She tells me the x-ray revealed what really happened. The piece of bone did not come from her femur, but from the back of her pelvis. She told me she was taken to the trauma center so the bone could be grafted back into place. I think, I'm glad I found the piece of bone and brought it to the hospital, but I kick myself for taking her to the local hospital and delaying her care. I was only given part of the story. My patient didn't just fall on the pipe with her leg; she sat on the pipe. She was actually impaled from the back of the leg all the way up to her butt bone. GROSS!!

What was is amazing is the fact that her husband lifted her off the pipe and carried her into the living room. This was quite a feat because she was no small lady. If we had found her impaled on the pipe, we would have treated her completely differently, and the bedpost would have gone to the hospital as well. Hindsight is 20/20; if only we could have that kind of hindsight when we are caring for our patients... until then, we'll do our best and hope we don't get blindsided.

Slingshot girl

Screaming is not a pleasant sound on a trauma call. There are different kinds of screams. There is the scream of fright, there is the scream of terror, there is the scream of joy, and there is the scream for help, and then there is the head injury scream. Let me tell you about this scream. Not many have heard this scream. Probably a good thing since you will never forget it. It burns into your memory the horrific nature of the event, and your memory will be seared by the event. The first time I heard this scream, I recognized the emergent nature of the call. I remembered my teachers telling me about this symptom and the dire condition of the patient.

This is the story.

Some teenagers are returning home from a football game. They are in a hurry to get home. They are driving with the windows down and are enjoying the evening fall air. The driver suddenly remembers that he has left something at the game and needs to turn around. Ahead is a wide section of road and he goes to whip a u-turn. The car behind him doesn't see him making this unorthodox move and accelerates to get around the erratic driver. The impact is horrific. The car hits the compact car in the back rear passenger side and sends the car spinning like a top. The centrifugal force is intense. The kids are tossed around like rag dolls in a box, but are saved by the seat belts, all but one, that is.

A young girl in the back seat is lifted up like a pellet in a slingshot and fired out the window.

The first scream is a scream of fright. The force of the twirling car ejects the young teen out the open window. She skips across the pavement like a rock across water, only each impact causes increasing injury. The young girl comes to rest at the side of the road and the screaming begins.

When her teenage friends come to their senses, they run to her side. They scream for help, another scream we mentioned earlier. The second scream.

One of the teens is the young girl's sister and she sees her sister's face covered in blood and her body contorting in pain. The third scream comes from the sister; the scream of horror. This is when we arrive. Hearing the screams, we know it is not going to be pretty. The hair rises on our necks, but we have no time to participate in all the screaming. We go to work. To save a life.

We see a small car in the middle of the road, empty. The teens that occupied it are standing on the side of the road about ten yards from the injured girl. They are at a loss at what to do for their friend. She is writhing on the road side, screaming that head injury scream. The fourth scream. The repetitive scream starts low and are more like a whimper, but escalates to a loud deep scream.

They stop for a few seconds, only to start all over again. This is a sign of building pressure in the brain, which is caused from the terrible head trauma. We quickly package our patient by placing her on a backboard and in a cervical collar. She is still screaming, that escalating scream.

Off we go, after calling in another ambulance to care for the other teens. I start IVs and oxygen and attempt to place a tube into her lungs to assist her breathing. The tube placement is unsuccessful due to the lack of paralytics, which we did not carry at the time. She is still screaming. A haunting scream of escalation then silence, then screaming, again and then silence.

Even as we arrive at the trauma center, she is still screaming. We tell the doctors the story and they move the girl into surgery.

The neurosurgeon will drill holes in her head to let out the pressure, and they will try to medicate her so she will not have brain damage, but the outcome is grim. The scream of joy will have to be for another call, because the young girl does not recover. She dies several days later, and the sounds of screams have been replaced with tears of sorrow. All for the sake of a backpack left at the game, and a young girl who didn't think she needed to wear a seat belt. The screams are not easy to forget, even for me. I think I now know why I don't like that screaming music that is all the rage. It brings back memories I'd rather forget. As a result, I am not a fan of screaming, understandably.

Suicide Pact *or* What is Death Like?

Teens are drama experts, and the teens we are going to introduce you to are future soap opera cast members.

"We are going to do something after school and we aren't telling you what it is, but it will be so cool."

"Everyone will know when we do it, and they will all be amazed. We are going to discover the truth of something everybody is fascinated with, and we are going to be the brave ones to find out what everyone is afraid of."

This is the story the two teens tell their friends.

GSW is how the call comes down. This means there has been a

gun shot wound of some kind and they need an ambulance. The police usually go in on these scenes to secure all weapons, and make it safe for us. It is very difficult to take care of patients while being shot at.

We are about five minutes away and the fire station is practically around the corner. They arrive before us and start working on the patients. The police frantically call us into the scene. As I walk into the chaos, a police officer is standing at the door watching the medics working. He has a shocked look on his face and seems a little confused by the scene.

What I see are two young men lying on top of each other. Blood is dripping from the face of the boy on top and the other young man has a bullet wound to the forehead. I ask if they are alive, one firefighter tells me the boy on the bottom has no pulse. I say he is dead- A declaration more to focus the firefighters, than as a statement of fact.

"What about the other youth? " I ask.

The boy moans and says his eyes hurt. Amazingly, he is alive. I notice that the boy has a bullet wound to both temples and his eyes are swollen shut. The gun was put to his temple and the bullet went through the back of the eyes, taking out both optic nerves but missing its intended target: the brain. He is now officially blind.

I look over at the police officer and ask him if he has secured the gun. He shakes his head no and the firefighter lifts the blind boy off the other boy. There sits the gun in the hand of the blind boy.

We quickly take it and give it to the officer. He has a look of dismay on his face. He knew he was supposed to secure the gun before we entered the scene. The headlines would have read, "Medics shot by blind teenager." (There I go with the headlines again, but I'm sure that one would have made the news.) We place the boy on a backboard and cervical collar and move him to the ambulance. The helicopter was called and we have to drive to the landing site, because there is no landing site in the residential community appropriate for the helicopter.

En route, my partner asks the young boy why he had done this terrible thing? He says that he and his friend were wondering what it was like to be dead, so they made a pact with each other to commit suicide. This is what the young men were bragging to their friends about at school. We put the young man on the helicopter and return to the scene to see if we can find more evidence to explain this bizarre situation. When I walk back into the house, the scene is very surreal. I notice the lead shell of the small caliber pistol protruding from the wall. There is a puddle of blood by a pillow and the dead young man is covered by a sheet. A video game with a skull on the box sits on the coffee table. I sense the complete waste of youth permeating the room. I leave the house with no new information. I walk up to the officer, who has interviewed several of the teens that have gathered to look and see what all the ruckus was all about. The officer tells me the gun was a small caliber pistol. The boy with the shot to the temple was the first to pull the trigger.

He told me the blind boy told the other boy not to do it because it hurt so much. The other boy must have been more determined in his pursuit of this undiscovered country called death. He put the gun to his forehead and with a small caliber weapon, this means his brains were scrambled by the bullet bouncing around in his skull. The parents arrive as we prepare to leave and the police give them the sad news. All I can think is that one boy knows what death is like and he won't be telling anyone, and the other boy will be blind for the rest of his life. He will wait a little longer in the dark before he finds out what death is like. The stupidity is beyond amazing.

Such a terrible loss of life, wasted on the pursuit of a knowledge better kept in the dark.

Midsummer's Night Nightmare by Sue N.

Drip… drip… drip, bright red blood dripping steadily. His life is slipping away before my eyes and there is nothing I can do. Too much trauma, there is way too much blood gushing from his nostrils and his mouth.

He is lying on the gravel shoulder of a two lane highway, dying. I am told he is only 18, but I can't tell from his mangled face if he is 18 or 58.

I can hear the young girl who was driving, crying in the background. Bystanders are holding on to her, trying to console her, trying to help in someway. She is the second patient. Not medically, as she survived the rollover without a scratch. Psychologically, she is a mess. Her life just took a bad turn.

"I was going way too fast around the corner. It's all my fault, oh my God, it's all my fault!" she chants her mantra.

Firefighters start an IV, not an easy job. His limbs are mangled, leaving few choices for access.

As they quietly work, I am standing at the doorway of my rescue truck, using the cab light to draw up the drugs needed for rapid sequence intubation. My patient can't protect his airway, it is up to me to do that for him. I need to intubate him, fast. He has an obvious head injury and I am worried I will not be able to visualize his vocal chords when it comes time to pass the endotracheal tube. So much blood, even with suctioning, I pray I will be able to see his necessary landmarks.

Over my shoulder, I keep an eye on the two firefighters who are carefully securing his crumbled body to a backboard. They are kneeling on the shoulder of the dark highway on a beautiful hot August night. It is one in the morning every few seconds, I ask,

"Are you okay... do you need help?"

At the same time, I am ever-so-carefully drawing up the drugs I need to push through an IV so his clenched teeth will relax.

Many eyes follow my every move. I feel an enormous responsibility for I am the only paramedic on scene.

Bystanders track the discussion between the volunteer fire personnel and me. I am very conscious of keeping my voice under control; inside I am scared and feel completely inadequate.

I know deep inside I am not going to be able to help this young man. There is way too much trauma.

Definitive help is way too far away, even with a helicopter waiting nearby. I am only a paramedic, there are no miracles tonight.

Later, I am handed his driver's license. Staring solemnly back at me is a handsome young man. Staring back at me is a dead man. He never made it to the trauma center.

There is no sleep for me that night. The next day, I search for answers, wondering who he was. I plug his name into search engines, look through newspapers for an article about the accident.

Two days later, as I drive home from work I see a handful of people putting up a roadside shrine. Compelled, I find a place to turn around and drive back.

Slowly, I approach the group. I am unsure how to precede, what to say?

I introduce myself and express my heartfelt sympathy.

It is the patient's twin sister and brother, along with some close friends.

They ask questions.

I answer their questions the best I can, maybe they can find some peace.

They have enough of a burden to bear. Mourning for a lost loved one is enough. Blame heals no one. By speaking with them, I find some peace too.

Each day I go to work, I pass his shrine, and each day on my way home, I am reminded of this Midsummer's night nightmare.

And always I am reminded how an impromptu summer drive to the beach turned horribly wrong, and how lives were forever changed.

And I wonder how long the "what ifs" will haunt us? The pain of memory is so very sharp, but time dulls this memory and I am able to go on. I hope they will too.

Mrs. Angel

My partner and I are driving to our next post following a call. I am not supposed to be driving, but since we just completed two calls back to back, I am driving to let my partner chart. 'Auto vs. Pedestrian' comes over the radio. Interest peaks, as I'm sure we will get the call. It's my call as I am up. We throw on the lights and put the pedal to the metal. Code 3!

As I approach the patient, I notice bystanders starting to gather on the sidewalk from the various apartment complexes in the area. The scene is a busy road that has a turn lane separating each direction. Mrs. Angel is in the middle of the road, surrounded by firefighters. I walk briskly up to her left side. She is unconscious and breathing erratically. Firefighters are ventilating her and preparing for intubation. I ask for a tourniquet so I can start an IV. I also listen for lung sounds. I ask the firefighter to give some squeezes to the Bag Valve Mask. The lungs are clear. One pupil is blown and the other constricted. Mrs. Angel has sustained massive head injuries. She also has numerous fractures to both femurs and pelvis. I could feel the bones in her elbow grinding as I positioned her arm for the IV. I get the IV and run in the fluids to increase her low blood pressure. Death is knocking at her door. She is in shock from the trauma. After confirming tube placement by listening to the lung sounds in both lungs, sometimes I step back from the scene just to get a better perspective.

Yes, your adrenaline rushes through you on calls like this. By stepping back and taking in the scene, I have often learned a lot. I take my eyes away from the patient for just a few seconds to look up the road. Traffic is blocked in both directions and I see a Nike running shoe lying in front of the first car in the long line of traffic. I look down and see the matching shoe on Mrs. Angel's left foot. She must have flown about 50 feet. I look to the left and then to the right. All I see is crowds of people lining the streets. Most have their hands to their faces as they weep and wipe away tears. They weep because it was tragic. For all they knew, they could have been Mrs. Angel. They could have been going out for a jog on a warm sunny afternoon only to be struck down a speeding SUV. It could have been them left for dead in the middle of the road. We could be working on them. Trying to save them with such traumatic injuries. It was just a brief moment in Mrs. Angel's life, but a moment that changed it forever.

After we prepared her for transport and transferred Mrs. Angel

to Life Flight, I looked at my partner and said, "She's dead."
Nothing was going to bring her back. She was a victim of a tragic
accident. She was at the wrong place at the wrong time. By the
looks of Mrs. Angel, she must have taken great care of herself. She
was very athletic and it showed in her youthful looks.

What makes this call difficult to recall is the fact that so many
other people don't take care of their bodies. Mrs. Angel was just
taking care of her body and all she got was a raw deal. I don't know
if she lived or died, but I do know, Mrs. Angel's life will never be
the same. May you fly to the heaven and leave your pain and
injuries behind.

Mr. Trendelenburg

Looking back to when I was seventeen years old, I was so
angered by how much my car insurance was at the time. Most of
my income went straight to maintaining a car and covering it with
insurance. Now that I turned twenty-five years old, I see why. Yes,
my rates went down but I'm a safer driver than when I was a teen.
Don't get me wrong, insurance companies are still expensive, but
the true reason for why they charge teenagers more for insurance is
simple, it comes down to risk. Teenagers take too many risks on the
roads. Not to mention that teens feel indestructible and invincible.
They feel like they own the roads and it's their right. It's really just
a privilege that can be taken away in the blink of an eye. Aggressive
drivers often end up learning firsthand about the many privileges

that can be taken away.

On a wet night along a two lane highway, a teenager is out for a drive in his parents' car. Little do they know as their son is speeding at speeds of 90 mph, double the legal limit. I am sure if they knew, he would be grounded. Punishment so bad, he would be stuck, riding a bike to school until he heads off to college. He would be the laughing stock of the senior class. If only this was the case, because he chose what was behind door number two, the gates of a metallic hell.

As I approach the scene of his MVA (motor vehicle accident), I see a single car that log rolled several times and crashed directly into a tree. The roof struck the tree first and then wrapped around the tree like a hot dog bun on a hotdog. A side view mirror was at the base of the tree and the other pointing towards the raining sky, still attached. On the other side of the tree, the front and rear bumpers were almost touching. Debris was everywhere and the tree was tipped at a 75° angle. The roots were exposed at the base.

Firefighters attempt to reach the driver inside. He was trapped and all I could hear were screams of pain. I look at my partner in surprise that someone survived this crash. I figured no one could survive the impact.

It takes about forty-five minutes before I can see my patient, Mr. Trendelenburg. He is still trapped, but I can see the top of his head and shoulders. Blood drips down from his forehead. Everything the firefighters does results in pain for Mr. Trendelenburg. Since it is going to take a few more minutes before we are going to extricate him, I find Mr. Trendelenburg's father. He is really upset, watching the fire department work to free his son. I tell the father, "Although the screams sound bad, and your son is in a lot of pain, this is the best thing we can hear right now. That means he is alive." I also tell him I don't know the extent of his injuries, but I do know he is conscious and aware of his surroundings as to what was going on. The father thanks me and I return to Mr. Trendelenburg.

At about the one hour mark of extrication time, Mr. Trendelenburg's legs are free. Fire states he has leg fractures. He is almost upside down in the car so we just glide him down the mangled roof and onto the backboard. He immediately sits up from supine and reaches for his legs as he states with agony, "My legs! Oh, my legs!" I put my hand on his chest and hold him down and tell him to be still. I look down at his legs and it is obvious he has broken both of his femurs, the largest bones in the human body. Both of his pant legs are bulging circumferentially just above the knees. As I look further down his legs, his feet are hanging off the sides of the backboard. His heels are pointing toward the sky and his toes towards the ground. A firefighter places his feet back on the backboard as Mr. Trendelenburg screams in pain. I cut off his pants to reveal the fractures I knew were present. We pull traction on his legs to help control internal bleeding, and the Life Flight crew starts a couple of IVs. While doing this, Mr. Trendelenburg went unconscious due to shock. We raise the foot of the stretcher and run fluids into the IVs. He regains consciousness and we get him onto the waiting helicopter.

The only thing that kept him alive was the way he was positioned in the vehicle after it crashed. Both of his femurs were crushed and resulted in massive internal hemorrhaging within his legs. On average, you can lose about 1-2 liters of blood from a single fracture to your femur. Mr. Trendelenburg had bilateral femur fractures and when his body only had about five liters of blood, you can see how this could be a problem. By having his legs above his head, this allowed for constant blood flow to continue to his brain and heart, which kept him alive. In any other position, he would have died on the scene. He was in perfect shock position, otherwise known as the Trendelenburg position. It saved his life so he could learn about aggressive driving the hard way. I just hope all his friends at school are smart enough to learn from somebody else's mistake. Months of surgeries and rehabilitation await Mr. Trendelenburg.

All drivers should ask one question "Is it worth it?" Stay off your phones, stop eating, doing makeup, shaving, speeding excessively and realize the risks you are taking. Everyone is affected in some form or another by your aggressive driving. Look out for yourself and cover your ass, because if not, someone will take you out. I personally want to go out on my own terms, don't you?

HOH

Our last few stories have been sad and melancholy, so I decided to tell a lighter story to finish this chapter.

We are called to a high mechanism car accident. A large Ford truck has had a head on collision with a larger truck, and there are injuries on the scene. As we pull up to the scene, the smaller of the two trucks is occupied by an elderly couple. My partner attends to the gentleman behind the wheel, and I go to the woman on the passenger side. We find the air bags have been deployed. Both of our patients are actually doing pretty good considering the damage. The air bags worked; but because of the age of our patients, we advise that they go to the hospital for evaluation. As a precaution, we put c-collars on them and move them to backboards for spinal immobilization.

They have no changes in their condition as we move them to the ambulance. Mrs. Ford tells me she has some heart problems as we move her first into the ambulance. They are not injured very bad, so we chose to take both of them in the ambulance together. As I attend Mrs. Ford, I tell her I will have to put some stickies on her to place her on my heart monitor. What I do not realize is as I am explaining this to her, her husband is being put into the ambulance and overhears my conversation.

He looks up at me and asks, "Why are you going to put hickeys on my wife?"

I look over to him a little stunned. I have never been accused of putting hickeys on my patients before. I notice there are hearing aids in his shirt pocket. I reach over and take them and put them in my patient's ears. I then tell him that I was not in the habit of putting hickeys on my patient, no matter how good-looking. The lady calls me a schmooze, as I explain to her husband that I was placing her on my heart monitor, not putting hickeys on her. We all have a good laugh as we go to the hospital. I ask Mr. Ford if he wanted some stickies? He looks at me with a questioning look. I smile and tell him if he wants to, he could borrow some stickies so he could put his own hickeys – I mean, stickies – on his wife. He thanks me and just laughed.

For your information, HOH is for "hard of hearing."

Chapter X
Chicken Breathers

Let me explain. Imagine walking into an apartment and several people of Hispanic origin are running around in a panic. They are speaking rapidly in Spanish and your broken Spanish is catching something about the mother in the back room not being able to breath. Suddenly, an elderly woman comes up to you and says, "Che can't breath, che can't breath." Yes, this is what I was told are "chicken breathers," now I understand. Shortness of breath plus Hispanic panic (I know, not a politically correct term, but bear with me) equals a patient with anxiety and shortness of breath.

This is very important to understand in treating these types of calls because if you can calm the Patients family, you're halfway there in bringing relief and comfort to the patient, let alone their family.

The following stories are going to address these 'chicken breathers', also known as SOBs (no, not short for son of a bitch. In

EMS, it is used for "shortness of breath," though there are times we think otherwise), also known as "I (panting) can't (gasping) breathe!" Call... 9... 11!

Mrs. Balloon

When a person is breathing at eight times a minute, and all you hear are little sounds of air moving over the vocal chords, it is pretty much a sign that someone can't breathe. The effort of gasping for air has long since passed and the condition called "hypoxia" (meaning not enough oxygen) has assaulted the brain cells to the point of demise. Cyanosis – blue skin – is not a healthy color. No make-up would match that skin tone.

Maybe it's Maybelline, or maybe she's just knocking on death's door.

Mrs. Balloon is being cared for by a young woman, who is holding her hand as we arrive. Mrs. Balloon is at the end of her cycle of breathing. Two breaks away from respiratory arrest. The young woman pats her hand and says, "You are going to be okay; the paramedics are here." The room is full of furniture, large stuffed recliners and couches. We ask the young woman to step out. Mrs. Balloon is blue. Cyanotic blue! We quickly pick her up and move her onto our gurney. She takes her last breath. A firefighter begins to breathe for her with a bad valve mask and oxygen.

He states it is hard to collapse the bag and difficult to get air into her lungs. Suddenly, he says, "There, that's better." This is a very ominous sign, a product of being too aggressive with the BVM. I think he popped her balloon. We place a tube into her lungs and continue breathing for her. Something is not right. Mrs. Balloon is blowing up like a balloon. Under her skin is air from her neck to her legs. We put needles in her chest to let the air out. We call it decompressing a Pneumothorax – air between the lungs and the chest. We pop Mrs. Balloon! We wouldn't want her to float away. On arrival at the hospital, the doctor asks why Mrs. Balloon is popped. We tell him she was inflated by a BVM and we didn't want her to float away. He looks at us with a confused expression then smiles. He then understands. Yes, we cannot have that since she is going to be buried six feet under. Mrs. Balloon was a chronic obstructive pulmonary disease patient that succumbed to the final stages of the disease. Yes, she has gone to that place all helium balloons go! Our heroic efforts were for naught. I don't think blowing her up like a balloon killed her but I know it definitely didn't help. I think smoking all those years finally caught up with her.

Mrs. Percolator

"My grandmother can't breathe and I want someone to do something about it. Sure, she is almost 90 years old and we have DNR orders and advance directives, but I love my grandmother and she shouldn't die like this." This is what we hear as we walk into the nursing home. The nurse meets us at the door and says the patient has DNR orders. DNR is often a death sentence in these facilities. We are now obligated to grant the wishes of the relatives because of the threat of law suits. It's a crazy system, but it's all we got. I walk into the room and lying in bed is an overweight woman trying with all her might to breathe. Each breath is a struggle. As she attempts each breath, the sound of water and foam gurgle up to her throat and pink frothy sputum drips from the edge of her mouth. The look of desperation is on her face and she is turning a bluish color on her lips and finger tips.

Her granddaughter walks in behind me and says she wants me to do everything I can do to save her. She is very distressed, I think it's inherited, the panic that is, her grandmother grabs my hand, but all she can do is struggle to breathe. I look at my partner and he knows we have to do something. He pulls out the airway kit and we medicate her and place a tube in her airway. I spend the remainder of the transport suctioning her airway. The doctors look at me as I turn her care over to them as if I am an evil pariah. What was I thinking trying to save this old woman. I should have just let her die. Which, by the way, is what they did after talking the family into pulling the tube? All my efforts were derided by the very doctors I look to for support and direction. My first lesson in a system that does not want to face the ethical questions of death and fiscal duty. Here I am caught in the middle again.

Mrs. Wheezes

Water on the lungs (we call crackles,) means a patient is drowning in their own fluids. The sound of whistling, we call wheezing, is a patient strangling themselves to death on a cellular level. The lungs have closed for business and the forced air makes a whistling sound. When we walk into a room, and you hear a patient wheezing with every breath, this is a bad sign. They are not whistling Dixie; they are trying to tell you that they can't breath. This is what happens when we walk into the home of Mrs. Wheezes. The house is decorated with breathing apparatus. No, not fireman equipment; nebulizers of all kinds, hand-held nebulizers, electric powered nebulizers and oxygen tanks, just to name a few.

Mrs. Wheezes has been at war with COPD – Chronic Obstructive Pulmonary Disease, and today she is losing. She is sitting on the couch with her hands on her knees, leaning forward, we call this the tripod position.

Each breath is a struggle. With broken words and about two words at a time, she tells us she awoke with a cold and that she has been puffing on her inhalers all day. She looks up to me and stops breathing. Her eye roll back in her head and Mrs. Wheezes has given up the battle to breathe. It is now my turn to step into the fray. I have been drafted into a war I unwittingly stepped into the middle of. The battle for Mrs. Wheezes' life is about to begin. I must breathe for her. I grab my equipment and place the BVM on her face and squeeze the bag that is full of life-giving oxygen.

Her lungs inflate with resistance. I am careful not to force the air in, but to provide gentle pressure. My partner starts an IV and gives Mrs. Wheezes epinephrine, a medicine we use that opens the lungs, enabling us to assist Mrs. Wheezes with each breath. We quickly place our patient on the gurney and place a tube in her airway. This is called intubation. It is essential for providing a secure airway. The short trip to the hospital is an effort in controlled breathing and medicating of Mrs. Wheezes. Our efforts are rewarded when Mrs. Wheezes begins to breathe on her own. The battle with COPD was won with our help- It is a sad fact of life that this disease will eventually win. The only cure that has been found is death.

Often, the cause of this disease is smoking, a slow smoky killer that is unforgiving and torturous, deceiving the smoker until the end.

Lesson: Stop smoking!

300 Pound PE

A woman is running around her apartment looking for her breathing medication. She has used up all the medication in the apartment and is frantic. She can't breathe and her panic is getting worse.

Her boyfriend calls 911. The fire department arrives and finds a 300 pound panicking woman that is running from room to room in her nightgown. Her boyfriend is standing in the living room telling the firefighters his girlfriend has asthma and that he has only been dating her for a couple of weeks. He tells them he doesn't know much about her medical condition. As I walk into this struggle for air, I see a firefighter chasing this panicking woman around the apartment with a bottle of oxygen and an oxygen mask, trying to get her to breathe the life-giving air. It is almost comical to see this desperate woman running around the apartment as she is, but I feel the intuition kicking in and tell the senior medic we need to get this woman down to the ambulance as soon as possible. The problem I foresee is that this woman who lives on the second floor of the complex will be very difficult to carry down the narrow stairwell. The firefighters finally corral our patient and she sits in the middle of the living room. Not a good idea.

She suddenly lies down and says, "I can't breathe." Then she sits up and says, "I can't breathe." My partner's eyes are the size of saucers. She has not seen this before and she also senses the need to get this patient to our ambulance, though for different reasons. She has only been a paramedic for six months and today she gets the baptism of fire. Our patient lies down for the last time. She stops breathing and the monitor goes flatline. What I have feared has come upon me. We have a 300 pound woman on the living room floor, dead weight in more ways than one. I look at the firefighter that I first suggested we move our patient to the ambulance, and ask him if their fire truck has a mega mover, a device invented just for this purpose. He says they do and he will call for more assistance to move her to our ambulance. I tell him this would be a good idea, as we begin our attempts and resuscitation.

My partner is at the head and tries to place a tube, but is not successful. Another paramedic steps in and gets the tube in place. We start the necessary medications and begin the difficult work of moving our patient. We transport our patient code 3 – lights and siren, to the hospital, where our patient is declared dead. We find out later that the patient was given birth control pills and was not told of the dangers of blood clots, especially for the overweight smoker. A blood clot was lodged in her lung and she suffocated. We were the witnesses of a painful death and we felt completely helpless. My partner later told me she had never seen anything like this, and was visibly shaken. This was the first death of a young person she had ever witnessed. I told her that with all our training and technology, sometimes there is nothing we can do. Sad to say, she will probably see this again, may it only give her more resolve to make a difference. From my brief time with her as a partner, I think she will be a great medic.

Chapter XI

The Thumpers

(not the Bambi kind)

We are trained in the art of emergency medicine. Like the painter, we have several mediums we use to create our masterpiece. The painter has acrylics, oils, and waters, and we have techniques, medications and technology. Paramedicine has evolved with the advancement of medicine, and has moved from the 'load and go' to the 'treat and transport.' In times past, emergency personnel had not had much more than a bottle of oxygen and a gurney. Today, we have monitors for the heart, the oxygen level, the sugar levels and even the amount of carbon dioxide we exhale.

Sometimes, all the wonderful tools replace the necessary knowledge that enables us to use those tools. And God forbid if these tools should fail us. We would be so lost and so would our patients. What is sad is that along with the technology comes the

"Trunk monkey." We are making the job so easy, 'even a cave man could do it.' Don't think, do! has become the mantra of making money in EMS. Quantity over quality. Sacrificing some to make more; more money that is.

We fire the experienced, so we can hire the inexperienced to save on that payroll check and gamble with the lives of our product (YOU!).

Dumbed-down medicine; It's cheaper and the only way you can make a killing in EMS. (Note the pun I won't apologize for this one)

"A fool and his money are soon parted," and we are the ones to exploit it. A little sarcasm here; Someone call 911!

Socialized medicine isn't the answer, either! Our communist and socialist rivals have proved this. In Latin, *res ipsa loquitur* (the thing speaks for itself) is very applicable to their outcomes.

The business of EMS should be people. The balance must be legislated, somehow. Greed must take a back seat when it comes to this noble profession. Caring and saving lives should be our mantra. So, Paramedics and EMT's don't forget!

Here are some stories that will help you remember.

Get back here! Something isn't right!

An elderly gentleman is having chest pain. He has chest pain that is called "Angina." He has pain on a regular basis, which he treats with nitroglycerin. But today, his medication isn't working and he needs to go to the emergency room. Our partners in crime have given him aspirin, nitro and started an IV (Intravenous line) and when we arrive, Mr. Wong is pain-free. They tell us Mr. Wong almost "passed out," before he called 911, and he is a little apprehensive. The hospital is only two miles away, but he wants to go by ambulance. His wife would take him, but she is scared to. So off we go. I get to drive today and my partner will take care of Mr. Wong. As we pull out of the driveway, my partner says Mr. Wong is starting to have chest pain again and she is going to give him some more "nitro." I keep an eye on her in my rearview mirror and watch the apprehensive look on her face as the patient looks to her for pain relief.

"Do you want to code 3, lights and siren?" I ask her.

She looks up at me and says, "Something is wrong with Mr. Wong, get back here." I pull off to the side of the road and step to the back of the ambulance. As I open the door, Mr. Wong is in a full body seizure. His head is turned and foam is coming out of his mouth. Something isn't right! I glance over to the monitor and notice Mr. Wong is in ventricular tachycardia, a heart rhythm that gives him no pulse and thus the seizure.

I reach over, and with my fist, thump him on the chest. He looks up at me as if I assaulted him and asks, "What happened?"

Well, "Sir, your heart wasn't working right and we had to give you a little jump start." My partner looks at me as if I'm an angel. I just smile, suggest some Lidocaine and tell her to repeat my performance if it happens again. Mr. Wong doesn't have his heart go awry again and he gets a fancy piece of technology implanted in his chest. His own personal "paramedic thumper". My partner learned the value of quick action and told me next time she calls me back not to beat on her patients. I think she learned something anyway?

Should I hit him again?

I have always been sensitive to my patient's medical conditions. I guess you develop a sense of how your patient is doing when you have done this long enough. One of my instructors taught me, from the beginning, to continually check your patients. (Thanks Jan.) Asking them how they are doing and to monitor their vital signs. I have been fortunate enough to be aware enough to intervene by taking life-saving measures. Our patients' conditions sometimes are not so simple. Mr. Scotch was a heavy drinker and was prone to all those diseases that come with slowly poisoning your body over the years. His liver was failing. His coordination was off, thus leading to a call to 911. Mr. Scotch was going out to the laundry room when he lost his balance and fell onto the boxes of laundry detergent.

He is now short of breath and can't get up. His ribs are hurting and my partner states she feels a rib that might be broken. We are not able to get a good "story" from Mr. Scotch due to his inebriated state. My partner decides we should go to the "trauma hospital." We put Mr. Scotch on a backboard and get an IV started on him. He admits to having had several shots of scotch and is a little drunk. It wasn't a revelation, but more an admission since we could smell the EtOH (short for alcohol) on his breath. The slurred speech is a sure sign he is feeling the "buzz." I put Mr. Scotch on the monitor and notice his heart rhythm is rapid and irregular. "Mr. Scotch, how are you feeling?"

"Not so good."

"Can I do anything for you?"

No answer.

I look over to the heart monitor and notice he has gone into v-tach (a bad rhythm). I shake him and he doesn't respond. I check his pulse; it is absent. I reach over and give him a "thump" on the chest, medically called a precordial thump. His heart rhythm changes and his pulse returns. Mr. Scotch looks up at me and says, "What happened?"

"You passed out," I tell him. I tell my partner what happened and she bumps it up to code 3 – lights and siren. Mr. Scotch has a concerned look on his face and asks what is happening. I tell him because he is passing out, we are going lights and siren to the hospital and I reassure him that we are going to get him the help he needs. His eyes roll back and his pulse stops. The heart monitor shows him in v-tach again. I am very concerned at this point about hitting a man on the chest with rib injuries.

I look up to my partner and ask, "Should I hit him again?" She looks at me with a confused look.

She must think I make it a habit of hitting my patients. I tell her he is in v-tach again. She says

"It worked the first time." I thump on his chest again and his heart begins to beat normally. I grab the Lidocaine and administrate the appropriate amount. He does not have another episode for the remainder of the trip. After Mr. Scotch was turned over to the doctors at the emergency room, my partner tells me I "really shouldn't beat on trauma patients." I feel a lecture coming, but she just smiles and shakes her head. "You saved Mr. Scotch," she tells me. Sometimes, that which is a little unorthodox is what you have to do. I was a little conflicted at first, but I have learned since then that this is the nature of EMS. I had faced down death and won. It gave me confidence, but I knew this was not always the way it would be.

Mr. Corvette

When metal meets wood, usually metal wins, but when that metal is a corvette and the tree is two feet thick, metal loses. Mr. Corvette was cruising on a curved road when he looks up and two headlights are in his lane. He swerves and takes on a two-foot tall tree. The violent collision wakes the neighborhood. A man runs out with a fire extinguisher and puts out the flames coming from the hood of the corvette. Another bystander pulls Mr. Corvette's girlfriend from the passenger side and helps her get away from the fire and smoke. Mr. Corvette is trapped with a broken leg and a broken forearm. Another neighbor calls 911. The fire is usefully extinguished with the arrival of another neighbor's fire extinguisher. We get the call. High mechanism accident- car into a tree with flames showing. The call is nearby and we arrive first.

Mr. Corvette is coughing and screaming for help and his girlfriend is lying in the neighbor's driveway. She is very short of breath and struggling to breathe. I call for a second ambulance and begin to attend to Mr. Corvette's girlfriend. I place a c-collar on her and put her on a backboard. A second ambulance arrives and I turn the care of Mr. Corvette's girlfriend over to them. I tell them she may have paper bag syndrome, a condition where the lungs pop from the sudden impact. They tell me that they're going to put her on the helicopter and will tell the flight nurse.

I go to care for Mr. Corvette. The firefighters arrive and start the extrication process. The firefighters cut the roof off of the corvette and open the car like a can opener. It was a beautiful car and now it's a pile of scrap metal. Mr. Corvette is alert and aware of everything that is happening to him. He seems to be very indifferent. He says, "Just get me out of here. I have broken bones before while skiing, this is nothing." As we pull Mr. Corvette out onto a backboard, we notice he has a fractured ankle and a broken upper leg. He is holding his wrist, which also looks broken. Mr. Corvette is hurting, but everything else looks intact. We load him into the ambulance and off we go.

Mr. Corvette seems to be carrying on an intelligent enough conversation about skiing when he suddenly passes out. I have him on the heart monitor and glance over to see he is flat-lining. I thump him on the chest and he wakes up.

"Did I pass out?" he asks.

"Yes," I say with a little concern and questioning tone in my voice. This is when Mr. Corvette decides to inform us that he has a pacemaker and that when he was a kid, he had an enlarged heart. Mr. Corvette is only in his mid-thirties and this seems odd. He explains he had an infection in his heart as a youth and then proceeds to pass out again. Another thump. He wakes up again. We consider putting him on our pacer, but pull into the emergency entrance before we get a chance. We quickly unload Mr. Corvette and wheel him into the trauma room. The doctors gather around and Mr. Corvette smiles at them as they examine him for life-threatening injuries. They find only broken bones. Most of the trauma doctors leave the room. We tell the one remaining doctor that his pacer stopped working while en route and his heart has stopped. The doctor looks at us with incredulity. He doesn't believe us.

The paramedics who saves Mr. Corvette by beating on his chest; Yeah right! We take our equipment and gurney out to the hall like rejected dogs with our tails between our legs. The firefighter who had ridden along shook his head and couldn't believe the doctors didn't believe us.

As we stood in the hall discussing what we had done, we hear the doctor yell, "Get the paddles, his heart has stopped."

We look in and in a simultaneous voice yell, "Thump him!" The doctor reaches over and thumps him. Mr. Corvette is revived, yet again. Now the doctors believe us. We feel vindicated. Mr. Corvette will get a new pacemaker and we will have a new story to tell our fellow paramedics.

Maybe, just maybe, the next time we tell the doctors about our patients, they will believe us.

Catch the spirit

Altered level of consciousness is a term used often in EMS to describe our patients' mental condition. Usually, this is a result of a severe medical condition.

Sometimes, it even applies to me, but that doesn't count.

Today, we are called for a patient who has an altered level of consciousness. A forty-year-old man has passed out at work and they can't wake him up.

On our arrival, our patient is being cared for by some very competent firefighters. Mr. Feelwell is sitting up and holding his head. The firefighters, who are also paramedics, discover he is in atrial fibrillation. This is a common irregular heart rhythm that some people live with all their adult lives, but for Mr. Feelwell, this is a new condition that means if he stands up too fast, he passes out. His heart is going too fast for the rest of his body. He says he is feeling better, but we convince him with our powers of persuasion that he should go to the hospital.

On our way to the hospital, Mr. Feelwell looks up at me and says, "I don't feel so good," and passes out. I look at the heart monitor and lo and behold, he is in flatline and he has no pulse. Basically he is dead. His spirit is leaving his body. No, not today. I grab it (no I didn't see his spirit but it felt like it), and hit him on the chest. He jumps but just lies there.

I begin to start chest compressions and he begins to move. I look over at the monitor and his heart rhythm is in atrial fibrillation again. I did not learn this in school, in fact, school told me this doesn't happen. When someone's heart stops, they go into v-fib or v-tach, or asystole (from which they rarely recover, and they definitely do not respond to a precordial thump).

You don't go from an atrial fibrillation to flatline and you don't convert from flatline with a precordial thump.

I tell my partner, "His heart has started again," as we arrive at the hospital. His wife meets us at the door. I tell her it was close but he's still with us. She cries on my shoulder and tells me thank you. Then off she goes to see how her husband is doing. Mr. Feelwell will get some medications and go home. I will have to find another school dogma I can prove wrong. One thing I have learned in the practice of paramedicine is there is always an exception. My problem is I seem to be a magnet to such cases.

Just Cough

Technology has often displaced the old medicine. This is sometime a good thing and sometimes not. Chest pain is a frequent reason to call 911. To Mr. Cough, he has a reason. He has sustained chest pain, he is sweating, and he can't breathe. Oh, this has been getting worse for the last three hours. The statistics say that this is the standard for waiting during a heart attack. You're only suppose to wait fifteen minutes at the most, but everybody or almost everybody waits three hours.

This is called denial. We say denial is a river in Egypt, not a good idea when it comes to a heart attack. You see time is muscle. This is what they teach us in school, anyways. Mr. Cough is running out of heart muscle. He has waited until his heart is shutting down, thus the sweating, shortness of breath and the crushing chest pain. We arrive with our friends, the firefighters, struggling to get an IV in Mr. Cough. You see, Mr. Cough is going into a rhythm that makes the blood pressure drop and usually leads to death if not treated quickly. As we walk in, no luck on the IV.

My partner is a young fellow who has moved up the ladder quickly, and is very ambitious.

He looks at me and says, "Let's shock him." I smile and look at Mr. Cough from across the room. I can tell he is going downhill fast. He is gasping and holding his chest.

"Mr. Cough," I say, "Give me a big cough?" He coughs and his heart returns to a normal rhythm. I got this look from the firefighters and my partner, "Einstein has entered the building." I tell them they should be able to get an IV now. They do and his blood pressure comes up. It's an old trick called "Cough CPR." It's less traumatic than shocking your patient. My partner tells me jokingly I'm a killjoy. I smile and tell him I just wanted to deliver my patient from the hands of the electroshock therapists (*them*). It would have been so painful to experience their treatment. Defibrillation hurts, especially if you're awake. Mr. Cough coughed several times on the way to the hospital and was glad I taught him the trick. It saved him a lot of headaches and heartaches. He received some medication and a pacer implant. He will live a little longer as long as the pacer doesn't fail, or he forgets how to cough.

Chapter XII

Farewell to Arms and Legs and Extremities

No, this is not Hemmingway. The following stories address the loss of limbs. No, not tree limbs; human limbs: like arms, legs, fingers and toes. We often take for granted our ability to walk and use our hands. These stories will give us a new perspective on the blessing of having what we so often take for granted – our extremities.

You're pulling my leg

Prosthetics are man-made limbs that enable those who have lost an extremity to function normally. Some people lose their leg to a traumatic accident, some to the disease processes of diabetes. Some have their extremity removed by surgery due to the loss of circulation. Usually, someone who loses a limb has an opportunity to live a fulfilling and productive life. This brings us to the next call – another code 99. As we walk up to the farmhouse, a frantic woman is standing on the porch urging us to quickly come inside. The living room is heated by a woodstove and a pile of wood crowds the furniture. Our victim is face down in the corner of the room, up against the wall. In order to begin our life-saving efforts, we need to make room to work. I move some furniture from the center of the room and my partner and a firefighter sets our equipment down on a recliner chair.

The man is not breathing and does not have a pulse. His wife tells us he just got out of the hospital and that he has not been feeling well all day. He was putting wood in the stove when he collapsed and she called 911. I roll the man over and see a 'blue' face and vomit in the corners of his mouth. I grab him by the shoulders and tell my partner to take his legs. As we lift, I see the expression of dismay and fright come onto his face as he hands me our victim's leg.

"What should I do with this?" he asks. I quickly tell him to set it aside and take his stump. We are able to move him to the center of the room. We begin our reconstruction as another ambulance arrives. He is rushed off to the hospital and we return to the station. My partner looks a little unnerved and I recognized this was his first "code 99." I comforted him and told him he did a great job.

The next day, the firefighter came up to me and asked, "How did our gentleman do yesterday?" "Well, they got him to the hospital and the doctor was very concerned about his leg." I told him. The doctor declared him DOA. The young firefighter was confused until I told him that I was "pulling his leg." The poor firefighter became the butt of the jokes around the station. We even went so far as to get him an award that made him an honorary "leg puller."

Lathe Man

Putting off retirement has been all the rage. Everyone is doing it. If you can still work, why stop.

My story is a good example of the importance of retirement. Mr. Shouldov had worked in a metal shop all his employable life. He had learned how to use a metal lathe and was a very valuable asset to his employers. He made pretty good money and enjoyed the work. He could set up a template and turn out metal parts as fast as any trained in his skill.

The call came down as an industrial injury, code three. We arrive as the firefighters are calling for a helicopter. I walk into a machine shop that is stacked high with metal and machines. It is a maze of projects and metal flakes.

The man is sitting on a chair holding his bloody arm. "I was going to retire in eight days," is a phrase he repeats over and over as he tries not to scream in pain. He is pale and sweating. Mr. Shouldov has become a victim of poor design and unsafe working conditions, all preventable, but too expensive to address. As the firefighters attend to his injury and prepare him for a helicopter ride, I go into my Sherlock Holmes mode. What was it that caused this injury?

I follow the trail of blood through the maze of machines and find the monster that assaulted my patient. Blood is sprayed on the ceiling. Blood is splattered over the machine and the metal rod that protrudes from the mouth of the lathe drips with blood. The machine was designed to turn the rod, which was supported by a sawhorse-like stand at 2400 rotations a minute. For some reason, the rod came off the support device and bent. This was the weapon of assault. Turning at 2400 times a minute meant the rod that was about an inch in diameter hit Mr. Shouldov's arm about forty times a second. His instinct was to reach up and turn off the machine, but the 'off switch' was located where his arm should not have been. This is how you get an aluminum rod to cut off your arm. Not something, I would recommend, especially if you are planning to retire in eight days. They couldn't save the arm and Mr. Shouldov did retire a little sooner than he planned. He is now learning how to write with his left hand.

Mr. Maximus

As we pulled into Starbucks to get that stimulant that keeps us all going, a call blares over the radio.

"Man hit by MAX (a local passenger train)."

Only two blocks away.

So much for the coffee. We switch on the lights and siren and arrive first on scene. The MAX driver is frantically waving us over to the other side of the train. The train is at a complete stop and parallel to the platform. From a distance, I see an arm protruding from between the tracks.

Looks like another dead man, I surmise. I walk up to the extremity to evaluate the situation.

I look at the arm from different angles expecting a disembodied limb.

I have to find the body connected to that arm, I think.

I walk over to the arm and take it by the wrist.

Kind of stupid I know, but you have to check.

His hand suddenly grabs my hand, my heart leaps from my chest, I holler out, "It's alive!"

The movie Young Frankenstein comes to mind, but it's not raining and there are no thunder bolts or lightning flashes and I am scared to death. Young Frankenstein was a comedy, this wasn't funny!

I wouldn't be surprised if I looked down and discovered I'd pissed on myself.

Luckily, this wasn't the case. My pants were dry and the arm was attached to a man. Reanimation will have to remain on the silver screen for the time being.

I look down, and under the platform, I see "Mr. Maximus." His legs are pinned between the tracks and the platform. His left leg is gone and his right leg is mangled, but he is alive. Later, we are told "Mr. Maximus" was very hard of hearing and couldn't hear the MAX train coming. He stepped out in front of the train as it was pulling into the station. He was hit and thrown forward onto the tracks. The train could not stop and ran over his legs, severing one and trapping the other. When it finally stopped, it was about 20 feet from the point of impact.

The train was only going at 8 mph, but you can't stop a train fast enough.

What prevented his demise was the space designed into the platform that the remainder of his body came to rest in.

I guess an 8 mph train hurts just as bad as an 80 mph train, only slower.

It took over an hour to get Mr. Maximus out from under the train. We had the helicopter on the ground when we finally extracted him and his leg. We loaded him into the helicopter with his leg, and off he went to get his hearing aids checked. The legs were a lost cause, too mangled.

Don't mess with trains; even if you can't hear, you will lose.

Fingers and Toes

I considered putting this story under the "Crazies" chapter, but it fit so well here that I decided to tell it here, so here goes.(I know to many 'here's', but what the hell; sometimes I wish I wasn't here).

We are called to stage for a man who is off his medication and has a history of self-mutilation. Usually, he just cuts himself with a razor on his arms or legs so he can feel the endorphins of the pain.

This is his form of self-medicating. Today, he has chosen a different weapon. An axe! The problem with the axe is it is not sharp enough to leave the small lacerations that are necessary. The pain of razor blade cuts just won't release the endorphins any more. He is going to have to do something a little more extreme.

When you swing an axe, what can you cut on yourself? Not your fingers(the axe handle is too long) and not your arms, they are too busy holding the axe. Mr. Paine decided he didn't like the way his feet looked so he chose to modify them with radical surgery.

We get the call. "Code 1" and stage, man with an axe. The police arrive and call us in. As we pull up to the house, we see several officers standing outside. Their expressions are of disgust as they point us to the back. The patient is sitting on a chair on the back porch and towels are wrapped around his feet, the towels are soaked in blood. Mr. Paine has this euphoric expression on his face, almost orgasmic. He has found the pain he was looking for in an axe. The police tell us that the axe is in the shed along with the body parts he has lobbed off. The man just moans and smiles.

He is un-bathed and smells of sweat and feces. My partner removes the towel to see if the bleeding is controlled. To his horror, every toe has been hacked off on both feet. We prepare our patient by starting an IV to help him replace some of the fluids he has lost. We help him to our gurney. My partner tells me I need to find his toes and put them in a protective bag. I go to the shed. I follow his bloody footprints to the place Mr. Paine found the courage to perform his self-mutilation. As I gather up his toes, an officer comes up to me and tells me I am contaminating his crime scene. I hold up to him a hand full of toes, with the blood of my patient covering my fingers. I tell him I am getting the rest of my patient. He looks at me with disgust and tells me to proceed. I carry the toes to the ambulance and my partner asks me if I got them all. I was so distracted by the horror of the situation, I didn't even think of counting them. To my dismay, the thought of having more toes in my hand than ten flashed into my mind. I mean, it was a crime scene according to the officer. Fortunately, I had the right number and they were transported to the hospital to be reattached, or maybe not. You see, if they attach them, who is to say he wouldn't do it again. Thank God we are just the messengers (carriers of fingers and toes) and don't have to make that decision. I hope Mr. Paine gets some psychological help because that was just crazy. No matter how much euphoria one might get from such actions, eventually you're going to run out of extremities.

Chapter XIII

An Apology and an Excuse

A writer must teach himself that the basest of all things is to be afraid.
                                            William Faulkner (1897-1962)

We boldly write our stories knowing some will be offended, some entertained and some inspired.

Every author in some degree portrays himself in his works, even if it be against his will.

Johann Wolfgang Von Goethe (1749-1832)

We have by some unknown force unwilling written our stories because they need to be told.

Why did I write? What sin to me unknown dipped me in ink, my parents, or my own?

Alexander Pope (1688-1744)

God forgive us! For we have sinned.

We quote these writers, not because we count ourselves

amongst them, but in great fear and respect, we look to them as an inspiration for our feeble endeavor.

We open a door to the candy-coated, Hollywood-ized propaganda world of emergency medicine and reveal the true and sometimes not-so-glorious world of EMS.

Not to be bitter, but to be real. Reality is not always pretty. It is not always ugly, either. It is somewhere in the middle. It has been described as moments of sheer terror followed by hours of mind-numbing boredom. This pendulum swing is not rhythmic. It has no pattern. No statistics or averages can predict its swing. It cannot be reduced by analysis, no matter how much the statistician would wish it to be. EMS is more in the realm of the theory of Chaos.

EMS must always be about saving people, whether you get recognition or honor or money or anything. Every paramedic and EMT must resist those who wish to shift their focus off this very important reality.

EMS is _people!_ NOT money, NOT glory, NOT statistics, NOT accolades, NOT recognition, NOT anything else! (My wife said I got a little preachy here, but I hope you get my point.)

The beauty of this concept is it includes everyone and excludes no one. This is the nobility that drives and inspires and motivates all those called to this honored profession.

It is a calling. Not many wise, not many noble, but all, driven to help people, to answer the call for help. Beware of all those who would jade you, or pervert you from this very important focus.

REMEMBER THIS ALWAYS! EMS IS PEOPLE! *And paramedics are people, too.*

Sounds trite, but we must always be reminded and remind others in order to protect us from those who would have it otherwise.

Our therapy

As we began writing, we discovered something amazing. The stories came to us like a river rising. Each story reminded us of another and we could not write fast enough.

Another amazing thing happened, areas of pain and emotional wounds came to the surface. As we related our experiences as paramedics, the things we thought long forgotten took on a clearer perspective. We began to see the things that made us into the paramedics we are today. Some good and some not so good, but that is who we are. No one should stand in judgment of this experience. It just is! And we are who we are, and that is enough. Sure, we can get better or we could go the other way, but that is up to us. We are the masters of our own fate and this makes us feel pretty good. For so long, we have felt like slaves to a system, or ships driven by the storm of the business model.

No, we are paramedics, a noble profession, no matter what others say of us, we know and we are proud. Not arrogant, not full of ourselves, but confident, we can keep our focus and answer the call for help. This is our center and our hope. Let no one take it from you. Let them fire you, let them suspend you, but do not give up your dignity as a paramedic. If you do, they win. Don't let them win! By "them," I mean anybody who would take away your dignity.

Your Joy

We have addressed the bizarre, the crazy, the sad and the strange.

I hope you have enjoyed the rollercoaster ride.

We have exposed the underbelly of EMS and maybe even the back side a little. We have opened the door to dialogue (do not slam it in our face). Our joy is to save a life, or to a bring life into this world. Do not let your experience jade you when it is otherwise, but let it make you a stronger paramedic. Let each successful Code and rescue and delivery build in you the joy of this job. This is your joy. It is what the heart of paramedicine should be.

Farewell

We have found this to be a very rewarding undertaking. I recommend that all paramedics write down their observations and stories and tell someone. They are contagious, in a good way. The stories are an important reminder to us that EMS is people.

Thank you for reading our stories.

Craig and Justin

About the Authors

Craig Mills has been involved in EMS for 15 years and is an active Paramedic. He has been a volunteer firefighter and is presently full time on the front lines of EMS. His background includes a degree in Theology, a certificate of Paramedicine and a year at UTEP premed. Craig has been training paramedics for eight years. He works in a system that provides ambulance service to an urban and rural community. His passion is helping people and that has been best fulfilled in the field as a Paramedic. He hopes you have an entertaining and enlightening experience reading the stories and seeing into his world.

Justin Miller has been a paramedic for two years and has also worked in a rural and urban setting. He is active on the front lines of EMS and has begun collecting a plethora of stories and experiences that would turn any bodies psyche or head for that matter. His hobby and therapy is ice hockey where he takes out some of his aggression and frustrations. His passion is Paramedicine and he hopes you enjoy the stories as much as he has in writing them.